DARWIN'S
ORIGIN OF SPECIES

A BEGINNER'S GUIDE

DARWIN'S
ORIGIN OF SPECIES

A BEGINNER'S GUIDE

GEORGE MYERSON

Hodder & Stoughton

A MEMBER OF THE HODDER HEADLINE GROUP

For Richard Baughan, with grateful recognition

Orders: please contact Bookpoint Ltd, 78 Milton Park, Abingdon, Oxon OX14 4TD. Telephone: (44) 01235 400400, Fax: (44) 01235 400500. Lines are open from 9.00–6.00, Monday to Saturday, with a 24-hour message answering service. Email address: orders@bookpoint.co.uk

British Library Cataloguing in Publication Data
A catalogue record for this title is available from The British Library

ISBN 0 340 80186 7

First published 2001
Impression number   10 9 8 7 6 5 4 3 2 1
Year 2007 2006 2005 2004 2003 2002

Cover photo from Corbis Images.
Typeset by Transet Limited, Coventry, England.
Printed in Great Britain for Hodder & Stoughton Educational, a division of Hodder Headline Plc, 338 Euston Road, London NW1 3BH by Cox & Wyman, Reading, Berks.

CONTENTS

FOREWORD

Welcome to …

Hodder & Stoughton's Beginner's Guides to Great Works

… your window into the world of the big ideas!

This series brings home for you the classics of western and world thought. These are the guides to the books everyone wants to have read – the greatest moments in science and philosophy, theology and psychology, politics and history. Even in the age of the Internet, these are the books that keep their lasting appeal. As so much becomes ephemeral – the text message, e-mail, the season's hit that is forgotten in a few weeks – we have a deeper need of something more lasting. These are the books that connect the ages, shining the light of the past on the changing present, and expanding the horizons of the future.

However, the great works are not always the most immediately accessible. Though they speak to us directly, in flashes, they are also expressions of human experience and perceptions at its most complex. The purpose of these guides is to take you into the world of these books, so that they can speak directly to your experience.

WHAT COUNTS AS A GREAT WORK?

There is no fixed list of great works. Our aim is to offer as comprehensive and varied a selection as possible from among the books which include:

* **The key points of influence** on science, ethics, religious beliefs, political values, psychological understanding.

* The finest achievements of **the greatest authors**.

* The origins and climaxes in **the great movements** of thought and belief.

* The most provocative arguments, which have aroused **the strongest reactions**, including the most notorious as well as the most praised works.

* The high points of **intellectual style**, wit and persuasion.

READING THIS GUIDE

There are many ways to enjoy this book – whether you are thinking of reading the great work, or have tried and want some support, or have enjoyed it and want some help to clarify and express your reactions.

These guides will help you appreciate your chosen book if you are taking a course, or if you are following your own pathway.

What this guide offers

Each guide aims:

* To tell the whole story of the book, from its origins to its influence.

* To follow the book's argument in a careful and lively way.

* To explain the key terms and concepts.

* To bring in accessible examples.

* To provide further reading and wider questions to explore.

How to approach this guide

These guides are designed to be a coherent read, keeping you turning the pages from start to finish – maybe even in a sitting or two!

At the same time, the guide is also a reference work that you can consult repeatedly as you read the great work or after finishing a passage. To make both reading and consulting easy, the guides have:

* Boxes identifying where we are in the reading of the great work.

* Key quotations with page references to different editions.

* Explanations of key quotes.

Our everyday life is buzzing with messages that get shorter and more disposable every month. Through this guide, you can enter a more lasting dialogue of ideas.

George Myerson,
Series Editor

A NOTE ABOUT QUOTATIONS

There are a number of widely available editions of *The Origin of Species*. In this beginner's guide, each quotation is followed by the page reference for three of the main editions:

P – Penguin Classics edition, edited with an introduction by J.W. Burrow (Harmondsworth, Middlesex: 1968). This uses the text from the first edition of the *Origin*.

O – Oxford World's Classics edition, edited with an introduction by Gillian Beer (Oxford, 1996). This uses the second edition of the *Origin*.

ML – Modern Library edition published by Random House (New York, 1993). This uses the sixth edition of the *Origin*.

There are minor differences between the first and second edition texts, and some larger differences with the sixth edition. These have been briefly noted when directly relevant.

SPECIAL FEATURES: DARWIN'S *ORIGIN OF SPECIES*

This Beginner's Guide aims to bring to life the reading of this great work, and to put that reading in context. For this purpose, a number of special features are included in the text:

Key Quote Boxes: These give a touch more emphasis to the presentation of extracts that are being considered in more depth or are more central to the understanding of Darwin's arguments.

Key Passage Boxes: In a few cases, longer extracts have been given the heading 'key passage' for emphasis and to highlight the different nature of their place in our discussion.

Close-Up: A few short extracts have been given this heading to accompany an unusually detailed view of their language.

Each of the above features is often accompanied by a section headed '***Anatomy***' of the key quote or passage. In these anatomies, central phrases are picked out and explained in context.

Quotation Boxes: These frame passages from the original text that are woven into the guided reading.

Key Terms: Occasional key terms are picked out for special definitions under this heading.

Bullet points are used to give clear summaries of the progress of our account, and chapter boxes will enable you to locate easily the movement through Darwin's text.

We hope the result will be a flowing discussion that fills in difficult points for you without too much interruption.

A GREAT WORK: CHARLES DARWIN'S *ORIGIN OF SPECIES*

Published 1859, Charles Darwin's book is one of the fundamental achievements of the modern world:

* The *Origin* is the founding book of 'the theory of evolution'.

* In the *Origin*, Darwin launches two of the most influential concepts in modern science, which have also helped to shape modern politics and visions of society:

  The Struggle for Existence
  Natural Selection

  Ultimately, these concepts have reshaped our sense of the place of humanity in nature.

* The *Origin* presents a vivid way of looking at the natural world, at landscapes and bodies, a way of listening to bird-song and seeing rocks by the seashore.

Darwin called his book, 'one long argument'. The aim of this guide is to bring that argument to life.

INTRODUCTION: GLIMPSES OF DARWIN

The Origin of Species, which first appeared in 1859, revolutionized the understanding of the natural world and of the human condition. A huge amount of controversy has raged around the *Origin*, from its first year until now, and no doubt this will continue through the new millennium.

Yet an aura of quietness surrounds the author of this great work, and a distinctive quiet emanates from many of its most important pages. This is the peculiar quietness of **looking**, a deep and concentrated gaze that takes in the natural world from flies' eggs laid in the navels of horses to vast forest landscapes, from pigeons in a fancier's loft to fossil skeletons of the first mammal. This quality of looking is both deeply personal and rigorously scientific, both meticulously objective and endlessly inquiring. To look though the eyes of the *Origin* is to think the world afresh, as well as to perceive its inhabitants more clearly.

DARWIN LOOKING

Black specks

Let's begin with three pictures of a man looking. In Desmond and Moore's authoritative life of Darwin, we find Charles as a university student in Edinburgh in the spring of 1827, filling a notebook with details about life forms found by the seashore, '*the larvae of molluscs and sea-mats, and the stalked sea-pens*' (Desmond and Moore, pp. 37–8). Encouraged by the naturalist Robert Grant, Darwin then gave a talk on 27 March to the Natural History Society at Edinburgh University. On this first public occasion, he described his having observed that sea-mat's larvae were able to swim and '*that black specks in old oyster shells were leech eggs*'. This last observation captures something important about Darwin's way of looking, even at that early stage. First of all, he takes in those tiny dots: there is a limitless *interest* at work behind those eyes. But then he also deciphers a new relationship. What others have assumed was one organism, he sees – and shows – is actually an interaction between two different organisms. You might think this was just a static entity, this old oyster shell, but under Darwin's gaze it starts to unfold a story, about the relationship between the oyster and the leech.

White rocks

Here is a second glimpse of Darwin, slightly later on. Now he is travelling as scientific officer on the exploration ship called *The Beagle*. He has landed on St Jago, Cape Verde Islands, in January 1832. They are 300 miles off the coast of Africa. There, according to his biographers: '*He spotted something odd – a horizontal white band running through the rocks*' (Desmond and Moore, p.117). Those white rocks were above sea level, but made of ancient shells and coral. Had the sea fallen? Had the land risen? What had happened to the landscape? Why had the sea moved? Or why had the land moved? Under the influence of the famous geologist Charles Lyell, whose *Principles of Geology* he read on the voyage, and who became a friend back home, Darwin continued to ponder these questions for years, and addressed them in different writings. The answers kept changing, and they involved general questions about the history of the earth. Here we glimpse another quality of Darwin looking: in his eyes, the scenes are cues, they are questions that *need* a new answer.

One gets a sense of the depth of Darwin's look when one discovers him coming off *The Beagle* in October 1836 with notebooks on geology totalling 1,383 pages and on zoology comprising 368 pages. He also brought specimens – 1,529 specimens in spirits and 3,907 labelled fragments and dried specimens. All of these were to be carefully catalogued. By the time of this return, Darwin was already renowned for the specimens he had sent back to London and Cambridge from the voyage.

Our third glimpse of Darwin looking finds him finishing this catalogue of specimens. He has got to the last barnacle: '*You cannot think how delighted I feel at having finished* (Desmond and Moore, p. 329). This one organism is left: an '*illformed little monster*'. Instead of scribbling a note, and passing on with relief, Darwin, with his friend the botanist Joseph Hooker, spent a lot of time wondering about the last barnacle. Together they gave it its own unique name: *Arthrobalanus, Jointed Balanus* – but Darwin was not entirely happy with the label, even then. Provoked by this object, and by wider dialogues, Darwin went on to write a series of volumes about the classification of barnacles. This glimpse shows something important about the relationship between Darwin's look and his words. This is a man for whom language is constantly under pressure from the intensity of his gaze. He finds the given words inadequate. They are too sloppy to capture exactly what he has seen, including the thoughts that are always implicit in any Darwinian moment of vision.

A BRIEF OUTLINE OF THIS GUIDE

* Chapter 1: The overall coherence of the *Origin*: Darwin's way of looking at the natural world.

* Chapter 2: Darwin's life and times; major intellectual influences.

* Chapter 3: Key words in the Darwinian language, as established in Chapters I and II of the *Origin* covering the domesticated and natural worlds.

* Chapter 4: The Darwinian Explanation. A twofold account of the major Darwinian concepts: 'The Struggle for Existence' (*Origin* III) and 'Natural Selection' (*Origin* IV).

* Chapter 5: The Darwinian Gaps: How Darwin recognizes the scope and limits of his own theory; Darwin's principles of knowledge and their role in the success of his theory; 'Mr Darwin's Planet' (*Origin* VI–Conclusion).

* Chapter 6: The Impacts of the *Origin*: The changing and continuing course of the Darwinian revolution in understanding.

1 The Darwinian Gaze

The *Origin* is not only very detailed and full of local examples, but it is also a profoundly unified book. In order to understand the working of the particular parts, it is useful to have in advance a sense of the complex whole to which they belong. The aims of this chapter are:

* to give an account of the inner unifying principles of the *Origin*;

* to map the development and structure;

* to bring into focus in advance key passages on which the whole argument turns.

Our consistent theme will be 'the Darwinian gaze', the inner spirit of the *Origin*, at once deeply personal and essentially scientific.

THE DARWINIAN GAZE 1: 'When we look ...'

After a brief introduction, the first words of Chapter I of the *Origin* are about looking:

> CLOSE-UP
> **When we look to the individuals of the same variety or sub-variety of our older cultivated plants and animals.**
>
> cross reference P: 71; O: 8; ML: 24

This is not just any look. One might call it 'the Darwinian gaze', and it is, in many ways, the central theme of this most influential of great works, Darwin's *Origin of Species*. If any single book can make the claim, then the *Origin* has indeed created a new way of looking at the world, a way of looking that literally begins in this rather innocuous-seeming sentence. For here, in fact, we are being politely introduced to the first principle of the Darwinian gaze:

The Darwinian Gaze: First Principle
Keep looking until you see the individuals.

When you think about it, we do not normally 'look to the individuals' at all! Far from it! These 'older cultivated plants and animals' include cows and pigs, wheat and apples, pigeons and barley, and the last thing anyone does, left to themselves, is to contemplate the differences between one individual cow and another, let alone one individual plant and another. The closest we normally come to doing this is with pets, but even then it is usually only the owner who fondly regards their cat or budgie as a unique and identifiable individual! Yet this is the first principle of the properly Darwinian gaze – this way of seeing. Look to the individuals!

It is noteworthy that the later editions revised the wording of this sentence by substituting 'when you compare' for 'when we look to'! This seems to confirm, in a curious way, that those original words mattered to Darwin. Perhaps they later seemed slightly too personal a way to begin a discussion which had become so authoritative.

Every edition keeps the emphasis on 'the individuals'. But what about that famous title phrase, 'The origin of *species*'? Isn't this meant to be a book about 'types' of plants and animals? Here we are approaching the great irony of Darwin's *Origin*; that origin of species is to be found not at the level of general types at all, but down among the masses of the individuals, precisely where scientists and philosophers of nature had never really looked.

No one could write such words out of nothing. In many ways, the whole of Darwin's own individual nature is contained in these few words: 'When we look to the individuals …', with that tactful 'we' concealing, or setting aside, the personal quality. Clearly this is the voice of a man who has become so used to such a gaze that he takes it for granted. It seems commonplace to him, simply a natural beginning.

THE DARWINIAN GAZE 2: The virgin forest and the Indian mound

'*When we look …*': what does looking mean for Darwin? To answer that question it is necessary to examine the whole structure and logic of the *Origin*. Chapter I looks to the domesticated world because this is the area where such a way of looking is actually more comprehensible. After all, enthusiasts for pigeons, breeders of dogs and cows, fruit farmers and shepherds do all have ways of gazing that zoom in on particular individuals among their chosen types. Chapter II then takes the argument forward to the realm of nature, mainly with the negative purpose of challenging the established way of seeing and describing life forms in terms of fixed categories. Here one might say that the book could have been called 'the end of species', as much as the origin: its immediate impact is to insist upon a look which dislodges general categories, like species, from the centre of the language of nature.

In Chapters III and IV, Darwin presents his great explanations of life on earth: 'The struggle for existence' and 'natural selection'. Here is a central passage from Chapter III, one to which we will be referring repeatedly. It is worth quoting at some length, because it shows how Darwin's explanation embodies his way of looking, his gaze:

KEY PASSAGE

When we look at the plants and bushes clothing an entangled bank, *we are tempted to attribute their proportional numbers and kinds to what we call chance. But how false a view this is! Every one has heard that when an American forest is cut down, a very different vegetation springs up; but it has been observed that the trees now growing on the ancient Indian mounds, in the Southern United States, display the same beautiful* **diversity** *and proportion of kinds as the surrounding virgin forest. What a* **struggle** *between the several kinds of trees must have gone on*

> *during long centuries, each annually scattering its seed by the*
> *thousand; what* **war** *between insect and insect – between insects,*
> *snails, and other animals with birds and beasts of prey –* **all**
> **striving to increase,** *and all feeding on each other or on the trees*
> *or their seeds and seedlings, or on the other plants which first*
> *clothed the ground and thus checked the growth of the trees!*
> *Throw up a handful of feathers, and all must fall to the ground*
> *according to definite laws; but how simple is this problem*
> *compared to the action and reaction of the innumerable plants*
> *and animals which have determined, in the course of centuries,*
> *the proportional numbers and kinds of trees, now growing on the*
> *old Indian ruins! [Author's emphases]*
>
> cross reference P: 125–6; O: 62; ML: 102

Again, as at the start of his first chapter, we have the Darwinian gaze: '*When we look …*'. This time the gaze is moving over the natural world, rather than the domesticated realm which was introduced by that opening chapter.

Anatomy of the Key Passage
When we do look, we see three landscapes here:

o First, there is *'the entangled bank'*, a scene which gives the appearance of a random intricacy. Plants coil and twist over each other in an apparent chaos, at least an intricacy too tightly meshed for us to unravel it.

o Second, we see *a forest clearing*, with the appearance of exploding new growth. This clearing has been made in an ancient forest and new plants and trees are observed crowding into the gap in nature. Again, it seems a landscape of chaos, too dense and too alive to analyse it.

o But then we have a third scene: *'the Indian mounds'*, a human site that has been reclaimed by the original forest. On these mounds, the human and the natural come together: they bridge the divide which Darwin has used to set up his argument in the opening two chapters. The mounds belong both to the domestic realm and to the natural world. When you look at the mounds you see growing on them the same proportions of plants as in the virgin forest.

There is a logic here, which over time produces a consistent structure. The true act of *observation* is this comparing of the mounds and the virgin forest, because in this moment you see that all the incredible complexity of such landscapes cannot be just random. A structure reappears, over a long enough time. On the Indian mounds, the Darwinian gaze has found a resting point, a still centre amidst the exploding complexity of life. The virgin forest and the entangled bank have a logic within them – it is just too hard for us to see normally. However, if there were no fundamental process at work, then the trees on those Indian mounds would never match those in the virgin forest.

Here in Chapter III, Darwin is putting forward one of the key concepts of the book: the struggle for existence. It is 'struggle' which explains this intricate pattern of the bank, the forest and the mound. If you look hard enough, you begin to see the struggle – what was a static scene becomes animated, as it does in Darwin's actual writing – '*each annually scattering its seeds by the thousand*', '*what war between insect and insect … insects, snails and other animals*'. This can be called a second fundamental principle of the Darwinian gaze:

The Darwinian Gaze: Second Principle
Keep looking until you see the process.

If you were to turn back to the entangled bank, and keep looking, you would begin to see the same 'struggle for existence' at work. What appeared to be chaos will begin to tell a story, a story of competition.

This view of the forest is also a key passage about 'origin' itself. The virgin forest seems always to have been the same: 'virgin', as if it were nature's original landscape, fixed forever from the beginning. But the Indian mounds show it is not a fixed original landscape. This virgin forest is the outcome of a long, vigorous process of struggle. Here the gaze mutates into an 'explanation': the forest is not a fixed original state, but a process unravelling through deep time, over countless centuries. The Darwinian gaze reconstructs this long history – incredibly long in human terms – using the intricate scenes of the present for evidence.

So we have a double irony: 'The Origin of *Species*' looks to **individuals**; and 'The *Origin* of Species' turns out not to be a beginning at all, but a continuing **process**.

THE DARWINIAN GAZE 3: 'Even the slightest ...'

From 'the struggle' Darwin moves directly to his other main explanatory concept: '**natural selection**'. Chapter IV contains probably the key passage in Darwin's great book, the one which displaces the Genesis story of creation, and which establishes the whole vision in general terms. In Chapter III we have gazed upon the struggle unfolding through every scene, every corner of the earth. Chapter IV explains how 'natural selection' is the force which governs that struggle. Darwin has referred throughout the first three chapters to the idea of natural selection, by analogy with human selection in domestic breeds. Now he gives the core definition of this concept of natural selection in its own terms. Inevitably, the language which Darwin finds appropriate for his key concept is the language of the gaze. We meet natural selection as a look – '**daily and hourly scrutinising**' all the individuals in the natural world:

KEY PASSAGE

It may be said that **natural selection** *is daily and hourly scrutinising, throughout the world, every* **variation***, even the slightest; rejecting that which is bad, preserving and adding up all that is* **good***; silently and insensibly working whenever and wherever opportunity offers, at the* **improvement** *of each organic being in relation to its organic and inorganic conditions of life. We see nothing of these slow changes in progress, until the hand of time has marked the* **long lapse** *of ages.* [Author's emphases]

cross reference P: 133; O: 70; ML: 112

Anatomy of the Key Passage

o Natural selection is the concept which corresponds to the Darwinian gaze. We have had, throughout the first three chapters of the book, Darwin 'looks at' life forms and now here is natural selection 'scrutinizing' the organic world. All editions after the first added the word 'metaphorically' following '*It may be said*' As we will see, Darwin also used the word metaphorical to describe his concept of 'struggle for existence'. In his usage, this word seems to confirm the breadth of an idea.

o The Darwinian gaze – whether embodied in the scientist or in his master concept of natural selection – is a look for which '*every variation, even the slightest*' is seen, and counted. Variations are precisely those characteristics that differentiate one individual bloom from another, or one sub-variety of cow from another.

Natural selection itself embodies what can be called the third – and most important – principle of the Darwinian gaze:

> ### The Darwinian Gaze: Third Principle
> *You cannot afford to overlook any difference, however tiny.*
> *Never assume that a difference is too small to count.*

Natural selection is the key concept of the *Origin* – it is the process which turns that 'struggle for existence' into what is now called evolution. In fact, it is another irony of this great work that it does not really use the term '**evolution**'. The *Origin* is not about the making of species by evolution, it is about how natural selection works on individual differences. The argument of the *Origin* is about how slight individual differences become magnified over huge spaces of time to give the types which we might unthinkingly assume were *permanent* 'species'.

Instead of being seen as a vehicle of the theory of evolution, the *Origin* is better accounted for as the story of this Darwinian gaze. The gaze of the scientist is worthwhile because the individual differences that it sees *count* – and natural selection is the *process* which has made them count. In Darwin's *Origin of Species*, natural selection, not evolution, is the name for the process to which such differences are important.

Here in this key passage from Chapter IV, we find Darwin defining natural selection in terms of a kind of super-gaze. But what natural selection sees isn't the whole of time, past, present and future. Natural selection does not have the omniscient gaze of the Christian deity. On the contrary, natural selection is the close-up gaze of the naturalist, but on an inconceivable scale. Natural selection is the gaze for which every single individual difference might be important – a diffrence which gives an individual any advantage in its struggle with all the other organic beings surrounding it. That 'advantage' is the key, not just because it will enable the favoured ones to survive, but because they will multiply. Successful individuals are the ones which produce more offspring and, Darwin believes, these offspring are likely to inherit the advantageous characteristics that their parent(s) happened to be born with themselves.

GAZE AND LANGUAGE: THE STRUCTURE OF THE *ORIGIN*

The *Origin* introduced a coherent language into the discussion of the natural world, something even larger than a '**theory**'. We have encountered some of the language of Darwin's gaze: natural selection, variation, species, checks, existence, diversity, war. Some of these terms were passed down to him, but he has remade all of them together into the modern language of nature, and he has done so by returning again and again to moments of 'looking'. Through this book Darwin bequeathed far more than a beautiful theory – he bequeathed a language of looking, one which was able to grow beyond the inevitable limitations of his own explanations and insights.

Let us sum up the logic and drama of the *Origin* in terms of the principles of the Darwinian gaze:

Principle	Concept
1 Keep looking until you see the individuals.	Variability
2 Keep looking until you see the process.	Struggle for existence
3 Never assume any individual difference is too tiny to count in the process.	Natural selection

In terms of the Darwin's theory, this brings in the three key concepts:

* *Variability:* the tendency of individuals to differ within all forms of organic life.

* *Struggle for existence:* the competition between individuals and their types in every landscape, in every corner of life, to survive and to reproduce.

* *Natural selection:* the process by which small differences give an advantage that is likely to be passed on through the struggle. From these inherited advantages emerge the forms that we call 'species' and 'varieties' or 'sub-species'.

In retrospect, this can be called Darwin's theory of evolution: in *The Origin of Species*, the ideas emerge as the content of Darwin's gaze, his distinct vision of life on earth.

For Darwin, theory is part of looking, and not an after-effect. **The aim of Darwin's gaze isn't simply to observe, it is to explain.** Darwin's great explanations – struggle for existence and natural selection – emerge from the fundamental principles of the Darwinian gaze. *The Origin of Species* recognizes no division between 'facts' and 'theories', or observations and explanations. There is throughout **a way of looking**, which unifies fact and theory.

After Darwin has introduced his key concepts of struggle for existence and natural selection, he shifts the nature of his argument. He picks out a series of difficult cases and problem areas. Behind this approach there lies, as we will see, a sophisticated and subtle view of scientific knowledge. Darwin aims to show that while his account is necessarily incomplete, there are no instances which can be taken to refute it. He covers a huge territory, from questions of fertility and infertility to the enigma of instincts, from the nature of the geological record to the influences of geographical features.

Throughout this sequence of questions and arguments, Darwin maintains the concentration of his gaze over the natural world and its history. The book ends with an extraordinary vision of the planet as a whole. There is, he claims, 'a grandeur in this view of life':

QUOTATION

...whilst this planet has gone cycling on according to the fixed law of gravity, from so simple a beginning endless forms most beautiful and most wonderful, have been, are being, evolved.

cross reference **P:** 460; **O:** 396; **ML:** 649

Few books have ever so deserved such a grand finale!

The Author of the *Origin* – A Modern Seer

This chapter presents:

* the brief life of Darwin leading up to the *Origin*;

* the intellectual influences and sources of the *Origin*,

* the wider context of ideas at the time.

The major sources for the account of Darwin's life are Desmond and Moore's biography, together with Stephen Jay Gould's work, notably *Ever Since Darwin*, Daniel Dennet's *Darwin's Dangerous Idea*, and Steve Jones's *Almost Like a Whale: The Origin of Species Updated*. Major sources for the context include Charles Coulston Gillespie's classic study of *Genesis and Geology*.

THE GROWTH OF A SCIENTIST'S GAZE

Educating Darwin

Charles Darwin was born on 12 February 1809 to Robert Darwin and Susannah Wedgwood. Robert was himself the son of Erasmus Darwin, a writer of massive poems on natural history. The family background was Unitarian, that is, at the rationalistic pole of eighteenth-century culture, as opposed to the mystical or the conventionally religious. Robert was a doctor (in Shropshire), giving a further scientific thread in the history.

Darwin file: Education

At the age of eight Darwin went to school where he made up natural history and claimed to have a power of changing colours of flowers (Desmond and Moore, p. 13).

In 1825 he went to Edinburgh University to study medicine and met the naturalist and teacher Robert Grant.

On 21 November 1826 Darwin joined the Edinburgh University Natural History Society.

In 1828 he enrolled at Christ's Cambridge, not having qualified as a doctor. Now he was destined for a church career.

On 24 March 1830 he took his first round of exams, and passed: 'Heaven protect the beetles …!' (Desmond and Moore, p. 79).

In 1831 Darwin graduated with a BA.

The Beagle expedition

Perhaps the true *Origin* story begins when Charles sailed with Captain Fitzroy on *The Beagle* as scientific officer and philosophical companion. They set off on 27 December 1831. In January 1832, we find him lying in his cabin reading Lyell's *Principles of Geology*, already embarked on the intellectual exploration that would lead to the *Origin*. As we saw earlier, he was soon sending back specimens in huge numbers from South America. By 1834 when they had arrived at East Falkland, Darwin's fossil megatheriums and other specimens had been sent to London and seen at meetings of the British Association for the Advancement of Science in Cambridge. One friend wrote to him that his '*name is likely to be immortalized*' for the South American specimens.

An aura of intellectual excitement surrounds the experience of this voyage. On 5 February 1835, for example, we find him writing up notes on fossil megatherium bones and speculating about extinction. Like Lyell, Darwin was busy dismissing climatic catastrophes, including Noah's Flood, as explanations for the bulk of extinctions. He was looking for some other process.

On 15 September 1835, Darwin sighted Chatham Island, the first of the Galapagos Islands. They were now a week from Lima, in Peru. Here (Desmond and Moore, p. 172) he famously encountered some '*most curious*' birds – the Galapagos finches, which became an important example in the building-up of his new approach. The expedition returned to England in October 1836.

Soon Darwin was back in London and working over the materials assembled on the great voyage.

Darwin file: Metropolitan life

On 29 Saturday October 1836 Darwin meets with Lyell, Professor of Geology in Cambridge. Lyell introduces him to Richard Owen, new Hunterian Professor at the Royal College of Surgeons (Desmond and Moore, p. 201).

On 4 January 1837 Darwin gives a talk to the Geological Society on the coast of Chile (Desmond and Moore, p. 207). Lyell was present.

By March 1837 Darwin was in London and attending gatherings at the home of the computer pioneer Charles Babbage.

Meanwhile, in 1837, Darwin was working with John Gould, a taxidermist, on those 'curious birds', from the Galapagos. They decided these were all related finches, Darwin having previously assumed they might be more widely diverse. It turned out he had gathered specimens of 12 species making up a group. This becomes, in retrospect, an important turning-point in Darwin's approach, a moment when he begins to conceive of the true flexibility of life forms, their potential for 'variation' and modification.

'Darwin's delay'

And here we begin to encounter a key aspect of his life, which may be called 'Darwin's delay'. By the late 1830s, he had already developed many of the key ideas and arguments of the *Origin*. Yet he did not write that book until the late 1850s, and then apparently only under external pressure. So what happened? Was he simply over-conscientious? Or were there other reasons for 'Darwin's delay' – reasons which help in understanding the book that eventually emerged?

Over these years, there is a marked separation between the published and unpublished writings. In his personal notes, Darwin was forging ahead with his new theories.

Darwin file: Unpublished thoughts

In mid-July 1837, Darwin begins a secret notebook, his 'B' notebook. Brown-covered, this volume was on transmutation of species.

In February 1838, he begins the maroon 'C' notebook. This contains such key Darwinian phrases as '*in the course of ages ten thousand varieties*', and '*those alone preserved which are well adapted.*'

By 1838 the 'C' notebook follows the ideas even beyond the range of the *Origin*: '*Man in his arrogance thinks himself a great work, worthy of the interposition of a deity, more humble and I believe true to consider him created from animals.*'

In October 1838 'E' and 'N' notebooks appear, on transmutation and metaphysics, involving general principles like '*the mind is a function of the body*'.

Main source: Desmond and Moore, pp. 229–69

The notebooks are full of data, but in a way that already contains all kinds of theories and explanations arising within the presentation of the facts. Meanwhile, the published works tend to stop short of these explanations, restricting themselves to more limited hints and questions.

Darwin file: Beagle Publications

Journal of Researches into the Geology and Natural History of the Various Countries visited by H.M.S.Beagle (late May 1839 and revised 1845).

The Structure and Distribution of Coral Reefs (1842).

Geological Observations on the Volcanic Islands (1846).

However we can already see, in retrospect, the essence of the Darwinian gaze of the *Origin*, appearing even in the published works, though without the whole Darwinian picture. From the published account of the voyage, here is a central passage about the Galapagos islands:

KEY PASSAGE

The natural history of these islands is eminently curious, and well deserves attention. Most of the organic productions are aboriginal creations, found nowhere else; there is even a difference between the inhabitants of the different islands; yet all show a marked relationship with those of America ... Considering the small size of these islands, we feel the more astonished at the number of their aboriginal beings.

Voyage of 'The Beagle', Everyman Library, 1959, p. 363 and after

---

Anatomy of the Key Passage

o **The Darwinian language is already at work: '*a difference between
the inhabitants*'. Darwin's distinct kind of attention is already
picking out unobserved differences.**

o **At the same time, Darwin has started to turn round the concept
of '*origin*' itself: 'aboriginal creations' are life forms particular to
that specific place.**

The implication of this passage is that there cannot be a single
unified act of creation of, say, the finch species, if every Galapagos
island has its own distinct type of finch.

Darwin zooms in on these intriguing birds:

> *'The remaining land birds form a most singular group of
> finches, related to each other in the structure of their beaks,
> short tails, form of body, and plumage … The most curious
> fact is the perfect gradation in the size of the beaks in the
> different species of Geospiza, from one as large as that of a
> haw-finch to that of a chaffinch, and … even to that of a
> warbler.'*
>
> *Voyage of 'The Beagle'*

Here the key phrase is '*the most curious fact*'. Darwin's gaze settles
inquiringly on minute differences and, in truly seeing them, he has
to tell a whole history of change and development. Each kind of
finch has a slightly different beak, but that isn't the truly
extraordinary thing. The greatest surprise is finding that there is a
perfect series of these little beaks, a sequence that is '*insensibly
graduated*'. That is the quality which Darwin discerns in life forms
generally, and which he uses to displace the previous assumption of
fixed species or types. That word 'insensibly' also contains a true
Darwinian combination of the visible and invisible. Only from the
perspective of a special kind of gaze do such differences *count*, the
gaze of the *Origin* and of natural selection.

> *'Seeing this gradation and diversity of structure in one small, intimately related group of birds, one might really fancy that from an original paucity of birds in this archipelago, one species had been taken and modified for different ends.'*

'*Seeing this gradation … one might really fancy*': this heralds the drama of Darwin's gaze – the gaze of inquiry, the gaze which demands an explanation. But even so he was already moving much further in the unpublished notebooks than in such a passage of public explanation, where he brackets his thoughts with the word 'fancy'. For Darwin, there *was* one parent species of finch, and now there are several, but he is not yet going to commit himself to that line of public argument.

Meanwhile, in July 1838 he was writing notes for and against marriage. On 29 May 1839, Charles married Emma Wedgwood. Intriguingly, his future wife noted his character as '*perfectly sweet tempered, and possesses some minor qualities … such as not being fastidious, and being humane to animals*' (Desmond and Moore, p. 269). In the course of a sustained and sustaining life together, they had ten children. In light and darkness, family life was always central for Darwin. It may have been the death of their daughter Annie in 1851 which gave a certain darker tone to the writing of the *Origin*.

So by now, after *The Beagle* voyage, we have seen a Darwinian approach forming, in both published and unpublished writings. This approach already has at its centre a particular way of paying attention to differences. The story in these terms is about an intellectual approach being shaped by a personal history. That story then must also overlap with an account of intellectual influences, the education of a gaze, the informed gaze.

KEY INTELLECTUAL INFLUENCES

Charles Lyell

One major influence has already been noted, Lyell, the major early Victorian authority on geology. The megatherium and the finches re-provoked a question which Lyell had asked in his classic book, *Principles of Geology* 'whether **species** have a real and permanent **existence** in nature'. This is the very question which caused Darwin to 'look to' individuals, and varieties, all kinds of differences that would be beyond the reach of the category of 'species' if it were regarded as a once-for-all and fixed entity.

Darwin's gaze at the megatherium and the finches passes through time as well as space. He is seeing the past in the present and here again, Lyell is an educative and formative presence:

> *the ancient history of the globe ... The disregard of this important subject may be attributed to the general persuasion, that former revolutions of the earth were not brought about by causes* **now in operation.**
> [Author's emphasis]
> Lyell's *Principles*, p. 162 (Penguin edition, 1997)

The key word which Lyell contributes to the education of the Darwinian gaze is 'now': causes don't lie buried in the remote past, they are still at work, observably so: '*the amount and kind of results to which ordinary subterranean operations are **now** giving rise.*' [Author's emphasis]

One can compare these passages of Lyell directly with Darwin's use of the word 'now' in such key passages as his depiction of those '*trees now growing on the ancient Indian mounds*'. If you look to the present, you can discern the past.

Thomas Malthus

Lyell was a personal, as well as philosophical, influence on Darwin. Thomas Malthus is more remote personally, but if anything more direct intellectually. His famous *Essay on the Principle of Population* (1798) was a shaping force not only for Darwin himself, but for Victorian thought in general. Whereas Lyell helps to form Darwin's whole way of seeing, Malthus contributes to the formation of a central idea: the 'struggle for existence', out of which the whole Darwinian explanation grows.

In his autobiography, Darwin recalled:

> *In October 1838 … I happened to read for amusement Malthus on* Population, *and being well prepared to appreciate the struggle for existence which everywhere goes on from long continued observation of the habits of animals and plants, it at once struck me that under these circumstances favourable variations would tend to be preserved and unfavourable ones to be destroyed.*
>
> Gould, *Ever Since Darwin*, p. 21 (P. 1991 reissue)

Darwin found in Malthus a theory of human population pressures, and their effects, a theory which he interwove with his own observations of the natural world. Malthus states, as we shall see when we consider Darwin's 'struggle', that food supplies would be outpaced by human population. Food would add up; population would multiply. From this Darwin adapted his vision of the struggle for existence. Gould notes pointedly that '*When Darwin achieved this Malthusian insight, he was twenty-nine years old …*'

The publication of Darwin's idea in the *Origin* would not be available for another twenty years!

BECOMING THE AUTHOR OF THE *ORIGIN*

A murder!

By the late 1830s, Darwin was already deep into a phase of explaining his amassed panorama of the natural world. In 1839, he recorded that he was '*steadily collecting every sort of fact, which may throw light on the origin and variation of species*' (Desmond and Moore, p. 286). Clearly the project of the *Origin* was already formed, and the influences of thinkers such as Lyell and Malthus were already being absorbed into a distinctive new way of looking at the world.

Then, in June 1842, staying at Shrewsbury, Darwin wrote 35 pages on his evolutionary theory in pencil. The text includes such terms as 'natural selection' and the 'war of nature'. On 11 January 1844, Darwin met with his friend Joseph Hooker and began to reveal his theory, which he described as 'a murder': '*I think I have found out (here's presumption) the simple way by which species become exquisitely adapted to various ends.*' He then handed over a 189-page essay, which contained the fundamental ideas of the *Origin*. (Desmond and Moore, pp. 286–313).

Robert Chambers, *Vestiges of Creation* (1844)

One reason why Darwin was driven on was the appearance of rival theories. At the same time, the reception of these ideas may well have discouraged Darwin from publishing his even more radical theory. The most notorious pre-evolutionary work was the *Vestiges of Creation*, published anonymously by Robert Chambers. This work stated categorically that life forms must be the expression of a continuing process of development, not the outcome of a single moment of creation:

> *It is most interesting to observe into how small a field the whole of the mysteries of nature thus ultimately resolve themselves. The inorganic has one final comprehensive law – GRAVITATION. The organic, the other great department*

of mundane things, rests in like manner on one law –
DEVELOPMENT.

C. C. Gillespie, *Genesis and Geology* (1951) p.153 (Chambers edn, p. 362)

Chambers struggled to reconcile this developmental theory with his faith:

How can we suppose an immediate exertion of this creative
power at one time to produce zoophytes, another time to
add a few marine mollusks ... This would surely be to take
a very mean view of the Creative Power.

But contemporaries were scandalized. Darwin, meanwhile, quietly recorded that:

In my opinion, it has done excellent service in calling in this
country attention to the subject, and in removing prejudices.

Gillespie, *Genesis and Geology* p. 217

The ugly facts

In the 1850s, Darwin closed in on his great work. In 1851 Lyell recorded being confronted by Darwin with some 'ugly facts' about nature. The public declaration was underway!

Darwin file: Drafting the *Origin*

In 1856 Lyell was taken around Darwin's pigeon house at Down House and told the theory. Lyell tells him to publish before someone else does.

During 1856 Darwin writes the first and second chapters on 'artificial selection'.

In 1857 Darwin writes a key chapter on 'Struggle for Existence'.

On Christmas day, 1857 Darwin finishes his 'Hybridism' chapter.

Desmond and Moore, pp. 444–48

Darwin was a prolific author, but this one was a battle. He notes wryly in 1856 that '*I find to my sorrow it will run to quite a big book*'.

In 1857, while Darwin was well advanced with his plan, he received a famous letter from a naturalist called Alfred Russell Wallace, who had been sending material from the Far East. On 18 June 1858, Wallace's letter arrived with a 20-page sketch for a theory of variation and struggle. Darwin was now anxious to establish his own claims before he was overtaken by a less coherent version.

On 20 July 1858, Darwin begins what he called an 'abstract' on natural selection. This is the revision that becomes the book. By April 1859, he had revised and reduced this text to 155,000 words. Lyell sent the work to the publisher John Murray, first as *An Abstract of an Essay on the Origin of Species and Varieties through Natural Selection*. This later became *On The Origin of Species and Varieties by Means of Natural Selection*.

Then Darwin took out 'varieties' to give the famous title. On 2 November 1859, Murray sent a sample copy to Darwin. Soon the 1,250 print-run was sold out and more were on the way. Three editions soon followed.

Darwin, of course, continued to write, and think, prolifically, but *The Origin of Species* remains his greatest work. In 1871, he published *The Descent of Man*, followed in 1872 by *The Expression of Emotion in Man and Animals*. His last book was the extraordinary study of one of his lifelong favourite animals, *The Formation of Vegetable Mould, Through the Action of Worms* (1881):

Worms have played a more important part in the history of the world than most persons at first suppose.

Charles Darwin died on 19 April 1882 and was buried in Westminster Abbey.

The Darwinian Language 3

This chapter draws mainly upon Chapter I of the *Origin*, with further examples from Chapter II and key passages from later in the book . Its aims are:

* to identify and explain key terminology;

* to show how Darwin **uses** this terminology – especially to hold in one sweeping argument facts and theories, observations with explanations;

* to explain what is truly **radical** about this way of looking and arguing, in the perspective of western thought.

Throughout, we need to be aware how the Darwinian language corresponds to the Darwinian gaze. It is far more than a matter of a few particular words.

DARWIN'S KEY WORDS

The end of 'species'?

When Darwin's friend, the botanist Joseph Hooker, read the draft of what became the *Origin*, his immediate reaction was shaky. He felt that some fixed point was being lost. Hooker was not a believer in biblical creation.

> ORIGIN
> **Chapter 1**
> *'Variation under Domestication'*

He was not shaken by any loss of religious faith or security. On the contrary, he had long passed over to the evolutionary side of that argument. Nevertheless, as a scientist, he felt that some bearings were becoming lost:

I never felt so shaky about the species before.

Joseph Hooker (1856) (Desmond and Moore, p. 445)

What Hooker felt he was reading might be called the 'end of species', rahter than the 'origin'. Even though he had left behind the religious view of original creation, he realized that he still had a reassuring sense of 'species'. His biological world was built up of stable **categories**. It is these categories that Darwin was determined to disturb. This follows, as we have seen, from the principles of the Darwinian gaze – to keep looking until you see the individuals, rather than stopping at the types. Darwin begins his great work by forging a new terminology for his gaze.

'Individuals'

As we saw, the *Origin* begins by 'looking to' the world of domesticated animals and plants, and seeing there not just types but endless **individuals**. In taking this approach, Darwin also committed himself to altering the balances between important words in the language he was using.

KEY TERM

Individuals are the sites of difference within types.

This terminology is essential in laying out a host of examples:

CLOSE-UP

When we look to the individuals of the same variety or sub-variety of our older cultivated plants and animals, one of the first points which strikes us, is, that they generally differ much more from each other, than do the individuals of any one species or variety in a state of nature.

cross reference P: 71; O: 8; ML: 24

What 'strikes' us now is how individuals 'differ much more': this is a language of variability.

Using these words, Darwin is consistently advancing his way of 'looking to' the living world, and that little word 'to' is itself interesting. It suggests depending on, trusting, putting faith in. The argument begins with this moment of turning towards individuals. It is this look, and its idiom, which marks the end of 'species' in an old sense, fixed and unchanging categories, essences.

In the history of western thought, this is a radical move. Here, for counterpoint, is one of the West's founding authorities, the philosopher and naturalist Aristotle:

> *… we can never affirm of a subject what is in its nature individual and also numerically one.*
>
> Aristotle's *Categories II* (Loeb Edition)

Aristotle is laying down the rule that knowledge cannot reach down to the level of individuals. Knowledge is about types:

> *The name of the species called 'man' is asserted of each individual.*
>
> Aristotle's *Categories IV* (Loeb Edition)

Aristotle believed that we cannot really *understand* the individual except in its relation to the 'species'. To understand John is to recognize how he is an example of the species 'man'.

You can see this classical logic still at work in, for example, Thomas Malthus, himself one of the main influences on Darwin:

> *There are individual exceptions now as there have always been. But, as these exceptions do not appear to increase in number, it would surely be a very unphilosophical mode of arguing to infer, merely from the existence of an exception,*

*that the exception would, in time, become the rule, and the
rule the exception.*

Malthus, Essay on the *Principle of Population* (1970) p. 71 (Penguin edition)

For Malthus, following Aristotle, to argue philosophically *means* to
discount the 'individual exceptions' – to focus on the general rule.
But for Darwin, these individual differences need to be explained,
rather than ignored – that is the logic of Darwin's gaze. It is a look of
inquiry and a cue for explanation. Indeed, Darwin's theory is
precisely about how the exception would, in time, become the rule:
the once rare advantage gradually becomes the norm for the newly
adapted species.

So later in the book, we find Darwin surveying the globe in terms of
his theory:

QUOTATION

*Widely ranging species, abounding in individuals,
which have already triumphed over many competitors in
their own widely extended homes will have the best
chance of seizing on new places, when they spread into
new countries.* [Author's emphasis]

cross reference P: 347; O: 283; ML: 486

A successful species is simply one which abounds in individuals.
Darwin's gaze individualizes life on earth.

'Variability'

Darwin then needs a way of talking about the relationship between
individual differences and general types. He is not arguing that the
world consists only of individuals, with no types or categories being
useful. But he is looking towards a new balance in the relationship
between individuality and type.

So when Darwin looks to all these differing individuals, he is not rejecting general classifications. On the contrary, he is seeing the '**variability**' of each category. This is the language which bears fruit in the definition of natural selection as '*scrutinising ... every variation*'.

> **KEY TERM**
>
> Variability is the power of a general type (or species) to give rise to new and changing differences among individuals.

What Darwin needs to explain isn't why a species is fixed, but why it is variable:

CLOSE-UP

the most frequent cause of **variability** *may be attributed to the male and female reproductive systems having been affected prior to the act of conception.*

cross reference P: 72; O: 9; ML: 25 (altered wording)

As the word 'cause' shows, this is the language of explanation. But you have to *see* the variability in order to want to *know* the cause, so the explanation is really an aspect of the gaze.

'Inheritance'

The next key term is '**inheritance**', which in Darwin's argument becomes radically revised as a concept. Previously, inheritance belonged to the political and moral language of stability and unity but in Darwinian language, inheritance is all about differences.

> **KEY TERM**
>
> Inheritance is the process by which differences are passed on.

If we turn to Darwin's massive database, we find plenty of evidence to support this approach to inheritance:

QUOTATION

Seedlings from the same fruit, and the young of the same litter, sometimes differ considerably from each other, though both the young and the parents ... have apparently been exposed to exactly the same conditions of life.

cross reference P: 73; O: 10; ML: text is revised here

Again, we meet the Darwinian language of 'differ considerably': under his magnifying gaze, small distinctions grow significant, becoming considerable, literally worth considering. Every word comes to life as Darwin seeks to retune the language to harmonize with his way of looking. This is a look which magnifies the differences, as we have seen in the third principle of the gaze – no difference is too small to count.

Of course, there are endless differences among individuals within any life form as it appears in the world. There is a real danger of being overrun by details here, of seeing *too much*. Inheritance is also the term which enables Darwin to clarify what really matters:

CLOSE-UP

Any variation which is not inherited is unimportant for us. But the number and diversity of inheritance deviations of structure, both those of slight and those of considerable physiological importance, is endless.

cross reference P: 75; O: 12; ML: 31

Darwin introduces a look which suddenly magnifies a huge diversity of details. But this Darwinian gaze is not about magnifying everything indiscriminately. There are differences which don't matter. The ones which *do* matter are those which are inheritable.

Here, in his approach to inheritance, Darwin also sees beyond his capacity to explain – and knows it: '*The laws governing inheritance are quite unknown*' (cross reference **P**: 76; **O**: 13; **ML**: 31).

One can usefully contrast this admission with the approach of Darwin's contemporary, Chambers, who believes he has explained inheritance:

> **the fundamental form of organic being is a globule, having a new globule forming within itself, by which it is in time discharged, and which is again followed by another and another, in endless succession.**
>
> Chambers, *Vestiges,* p. 173 (Gillespie (1951) p. 156)

Had Darwin adopted such an ill-founded explanation, his work would now belong to the prehistory of modern science.

In fact, the outcome of this Darwinian gaze is the future science of life. In his questioning and self-questioning, Darwin was leaving the space for that future, which Steve Jones calls '*Genetics, the science of differences*' (Jones, *Almost Like a Whale* (1999) p. 47).

We can still hear the Darwinian language everywhere in the continuing discussion of this genetics, as when the eminent genetic theorist Richard Dawkins declares, '*We are interested in natural selection, and natural selection is differential survival of genes*' (*The Extended Phenotype*, (1982) p. 18).

'Species'

All of Darwin's other key words serve to give him essential leverage on the most famous of all the terms – **species**. Darwin sweeps across the field of domesticated animals and plants, observing the huge variety of individuals. An obvious rejoinder occurs to him: maybe there is something odd about these types of organism, these species:

QUOTATION

It has often been assumed that man has chosen for domestication animals and plants having an extraordinary inherent tendency to vary, and likewise to withstand diverse climates.

cross reference P: 79; O: 16; ML: 35

Darwin vigorously discards this old assumption, in a characteristic gesture of dismissal:

QUOTATION

I cannot doubt that if other animals and plants ... were taken from a state of nature, and could be made to breed for an equal number of generations under domestication, they would vary on average as largely as the **parent species** *of our domesticated productions have varied.* [Author's emphasis]

cross reference P: 79; O: 16; ML: 36

Pigs and barley, cows and apples do not belong to special species. They have no unique capacity to vary. It is simply that (*a*) we notice their differences more vividly and (*b*) we have exposed them to unusually careful human selection.

In refusing the previous explanation, that these are exceptional species, that most do not vary, Darwin is actually making a controversial claim, and taking his own hold upon that word 'species'. He is breaking the link between species and creation, a once and only beginning, substituting his own bonding of species with selection, an ongoing process. Within the domestic realm, man is the agent of selection; outside, there will be *natural* selection.

The word species is being redefined. In Darwinian, it refers not to a fixed and given type, but to a field of related variations, and where we draw the line will be a matter of judgement. In the following passage, from Chapter IX of the *Origin*, we can see how important this use of the term remains:

QUOTATION

It is all-important to remember that naturalists have no golden rule by which to distinguish species and varieties; they grant some little variability to each species, but when they meet with a somewhat greater amount of difference between any two forms, they rank both as species.

cross reference P: 305; O: 240; ML: 426

For an example of the non-Darwinian use of the word species, we can turn to Darwin's own mentor and inspiration, Lyell:

It usually happens that those species, both of the animal and vegetable kingdom, which have the greatest pliability of organization, those which are most capable of accommodating themselves to a great variety of new circumstances, are most serviceable to man.

Lyell, *Principles of Geology*, p. 202 (P, 1997)

Lyell is arguing precisely that we have chosen to domesticate '*those species that have the most flexible frames and constitutions*'. In effect, Lyell is giving to this central word species an authority, a stature, which it loses in the Darwinian language.

For Lyell, 'species' are closed units, some more variable, most less so. For Darwin, 'species' is an open-ended field, and there is always the same potential for variation between individuals. For the Darwinian gaze, the variation among individuals takes precedence over the clarity of the species, and in order to make this vision communicable Darwin has to find a new language for differences, the language which in our time is still spoken (and developed) by genetics.

'Races'

The term **race** belongs to the language of classification in which Darwin is attempting to intervene and comes into play as part of clarifying the general question about origins. We have, for example, several races of domestic pigeons:

CLOSE-UP

The doctrine of the origin of our several domestic races from several aboriginal stocks, has been carried to an absurd extreme by some authors.

cross reference P: 80; O: 17; ML: 38

Perhaps, others may speculate, one could explain the variety of races of pigeon by tracing them back to several origins. Darwin is keen to resist this explanation, since it is a supposed alternative to his theory in which a single species can become modified in diverse directions. The term race comes in handy as a more open-ended concept than species: it suggests, as Darwin uses it, groupings that share a common origin, but not necessarily a common future. Indeed the full title of Darwin's book reads: *On The Origin of Species by Means of Natural Selection: or the Preservation of Favoured Races in the Struggle for Life.*

The term races was used in a similar fashion by Lyell:

We find a striking disparity between individuals which we know to have descended from a common stock, and these newly-acquired peculiarities are regularly transmitted from one generation to another, constituting what are called races.

Lyell, op. cit., p. 186

In this usage, 'race' defines an emerging area of slight differences within a 'common stock'. So a race is less distinct than a species, for Lyell, but more than a variety.

In the spirit of Darwin's wider approach to language, the modern geneticist Steve Jones comments on this area of confusing terminologies:

> *The issue of who belongs where in the natural world can sometimes be sidestepped with 'varieties', 'races' or 'subspecies'. What these are often depends on who studies them.*

Jones, op. cit., p. 51

One person's 'race' is another's 'variety'. The overall thrust of the Darwinian approach is to undermine the status of such classifications, to render them provisional and ambiguous.

We have now seen how Darwin introduces several of his key words in the opening arguments of his first chapter, and have looked forward at how they work as part of a Darwinian language of nature. Next it is time to see how Darwin introduces, and establishes, the main way of *applying* this language. Although the *Origin* is famous as the source of a great 'theory', it is in fact animated by an endless series of specific examples.

THE PIGEON'S EYELID: A DARWINIAN EXAMPLE

There are subtle changes in the meanings of key words throughout Darwin's arguments. But Darwinian language is far more than a question of terminology. It is a whole way of putting things, corresponding to the wider way of looking at the world. Above all, 'Darwinian' is a language for bringing individual examples to life. This language is seen at its finest in the handling of the example with which he develops the argument about common origins and domestic diversity: the pigeon.

Proudly, Darwin establishes his personal credentials for discussing this particular example:

QUOTATION

I have associated with several eminent fanciers, and have been permitted to join two of the London Pigeon Clubs.

cross reference P: 82; O: 19; ML: 39

This is not a work of theory which is designed to provide short cuts around other people's specific expertise. Darwin's language is never more alive than when acknowledging a local field of expert knowledge:

KEY PASSAGE

The diversity of the breeds is something astonishing. ... The carrier, more especially the male bird, is also remarkable from the wonderful development of the carunculated skin about the head, and this is accompanied by greatly elongated eyelids, very large external orifices to the nostrils, and a wide gape of mouth.

cross reference P: 82; O: 19; ML: 40

Anatomy of the Key Passage

o Darwin wants the reader to become excited by the facts of 'diversity'.

o Success is a function of this 'diversity of the breeds', which allows 'wonderful development'.

o Under Darwin's gaze, '*greatly* elongated' can readily be used of the eyelid of a pigeon.

From here, Darwin turns his attention to one particular breed of pigeon:

> ### QUOTATION
> *The rock-pigeon is of a slaty-blue, and has a white rump ... the tail has a terminal dark bar, with the bases of the outer feathers edged with white; the wings have two black bars.*
>
> cross reference P: 85; O: 22; ML: 43

He describes his own experiments and expertise as a pigeon breeder:

> ### QUOTATION
> *I crossed some uniformly white fantails with some uniformly black barbs, and they produced mottled brown and black birds; these again I crossed together, and one grandchild of the pure white fantail and pure black barb was as beautiful a blue colour, with the white rump, double black wing-bar, and barred and white edged tail-feathers, as any wild rock-pigeon!*
>
> cross reference P: 85; O: 23; ML: 44

His point is that the original characteristics of that rock pigeon reappear in some individuals of other breeds. It is clear, therefore, that whatever this great diversity of pigeons, '*all have descended from the rock-pigeon*'.

The pigeon example prefigures key moments from later in the *Origin*. Most importantly, in Chapter IX, Darwin will consider 'the great class of mammals', to which we ourselves belong. He records with excitement that, '*one true mammal has been discovered in the new red sandstone at nearly the commencement of this great series*'. The rock-pigeon has diversified to give many breeds; this original mammal too has been modified to yield many descendants, including human beings. Both cases belong to Darwin's argument against '*the common view of the immutability of species*' (IX) and in defence of his own theory of 'slow and gradual modification'.

Darwin's pigeons caused a stir, which shows how effectively he had presented his examples using his new language. Richard Owen, Darwin's friend, was curator of the natural history section of the British Museum. He used public interest in these birds to persuade a committee of Parliament to set up the Natural History Museum as a separate establishment. Owen told the MPs that people were flocking to the British Museum demanding to be shown the diversity of pigeons: '*Let us see all these varieties*'(Desmond and Moore, p. 490). A new display was needed!

To understand why Darwin's discussion of pigeons had such an impact, one can compare it with an equivalent instance from Lyell, where he focuses on recent finds in Egyptian archaeology. Excavation has revealed mummified domestic animals. Lyell is interested in what he sees as the unchanging character of domestic animals:

> *Now such was the conformity of the whole* **species** *to those* **now** *living, that there was no difference, says Cuvier, between them than between the human mummies and the embalmed bodies of men of the present day.* [**Author's emphases**]

> Lyell, op. cit., p. 205

For Lyell, the key phrase is '*no more difference*'. The mummified ancient cat is, he claims, identical to our pets:

> *The cat, for example, has been carried over the whole earth … yet it has scarcely undergone any perceptible mutation, and is still the same animal which was held sacred by Egyptians.*

Again, the phrase that counts is '*scarcely ...any perceptible mutation*'. After Darwin, what is 'perceptible' has changed! Looking again, we would see differences where Lyell notices none.

SELECTION

Now we have seen both the opening sweep of the Darwinian language, and its vivid application to a specific example, all set in motion by the very first chapter of the book. In the aftermath of the pigeon example, Darwin underlines the key term, which we will need for a full understanding of the later stages.

Darwin sums up his account of the domesticated realm:

QUOTATION

The key is man's power of **accumulative selection:** *nature gives successive* **variations;** *man* **adds them up** *in certain directions.* [**Author's emphases**]

cross reference P: 90; O: 27; ML: 50

A pig breeder notices individual differences and selects certain individuals from which to breed. Over time, such a process produces a new species of pig:

QUOTATION

selection ... *its importance consists in the great effect produced by the accumulation in one direction, during successive generations, of differences absolutely inappreciable by an uneducated eye.* [**Author's emphasis**]

cross reference P: 91; O: 28; ML: 51

Typically, Darwin defines this selection in terms of the *eye*. The expert eye sees differences where the ordinary eye sees none. It is this educated eye that gives '**accumulative selection**'. But human selection is not simply a matter of conscious preferences. The process is too complex for that:

QUOTATION

At the present time, eminent breeders try by methodical **selection** ... *to make a new strain or sub-breed,* **superior** *to anything* **existing** *in the country. But, for our purpose, a kind of Selection, which may be called Unconscious, and which results from everyone trying to possess and breed from the best individual animals, is more important.* [Author's emphases]

cross reference P: 93; O: 30; ML: 54

Experts see more than they realize: their gaze is too deep for their own consciousness. It is this unconscious guidance that links human selection to the wider process which Darwin calls natural selection.

ORIGIN
Chapter 11
'Variation under Nature'

In his next step, Darwin turns his own language against the established view of nature:

QUOTATION

... every naturalist knows vaguely what he means when he speaks of a species. Generally the term includes the unknown element of a distinct act of creation.

cross reference P: 101; O: 38; ML: 65

He is now ready to replace this vague act of creation with his own theory of 'selection':

KEY PASSAGE

*No one supposes that all the individuals of the same species are cast in the very same mould. These individual differences are highly important to us, as they afford materials for **natural selection** to accumulate, in the same manner as man can accumulate in any given direction individual differences in his domesticated productions.* [**Author's emphasis**]

cross reference P: 102; O: 39; ML: 67

Anatomy of the Key Passage

- o **Darwin redefines 'species' in terms of the range of 'individuals'.**

- o **In this new usage, 'differences' count 'for us'.**

- o **The term 'natural selection' is introduced as a parallel with the previous account of human selection.**

- o **This is a theory of 'productions', belonging to an industrial age.**

Darwin is now on the way to his vision of *'natural selection ... daily and hourly scrutinising, throughout the world, every variation'*. In this Darwinian landscape, the 'entangled bank' is everywhere. For example, he surveys plant populations or floras:

QUOTATION

See what a surprising number of forms have been ranked by one botanist as good species, and by another as mere varieties.

cross reference P: 104; O: 41; ML: 71

Amidst the tangle, the old terminology collapses. Experts used to be the ones who decided that 'this' was a species and 'that' a mere variety. But no longer! The history of botany emerges as a confusion of categories:

QUOTATION

Mr H.C. Watson, to whom I lie under deep obligation for assistance of all kinds, has marked for me 182 British plants, which are generally considered as varieties, but which have all been ranked by botanists as species.

cross reference **P**: 104; **O**: 41; **ML**: 71

Take, he goes on, the everyday case of the primula and the cowslip. Surely they are different species?

QUOTATION

These plants differ considerably in appearance; they have a different flavour and emit a different odour; they flower at slightly different periods; they grow in somewhat different stations.

cross reference **P**: 105; **O**: 42; **ML**: omits

Yet, Darwin argues, the two forms are in fact tangled together by many connections:

QUOTATION

We could hardly wish for better evidence of the two forms being **specifically distinct.** *On the other hand, they are united by many intermediate links … there is an overwhelming amount of experimental evidence, showing that they descend from common parents, and consequently must be ranked as varieties.* [**Author's emphasis**]

cross reference **P**: 105–6; **O**: 42

Darwin denies that there is anything specifically distinct about the primula and the cowslip: there is no species division, and no special division either! The old idea was that these plants were clearly separate; the new idea is a vision of entangled life forms:

QUOTATION

Hence I look at individual differences, though of small interest to the systematist, as of high importance for us, as being the first step towards such slight varieties as are barely thought worth recording in works on natural history ... steps to more strongly marked and more permanent varieties ... leading to sub-species, and to species.

cross reference **P**: 107; **O**: 44; **ML**: 77

Over the gateway to natural selection, hangs the phase: 'hence I look at ...'.

4 The Darwinian Explanation

The Darwinian gaze and the accompanying Darwinian language are not ends in themselves, as they might have been for a novelist or a philosopher. The aim of the *Origin* is to use this gaze and its language to present and justify a new explanation of how the earth comes to be inhabited by the life forms we observe upon it. In this chapter of this guide, we focus on the twin concepts of this Darwinian explanation: the struggle for existence and natural selection'.

The aims of this chapter are:

* to present Darwin's explanation of how life develops;

* to show the interrelationship of his two great concepts of 'struggle' and 'selection';

* to show how Darwin presents his ideas so persuasively.

EXPLANATION 1: THE 'STRUGGLE' ARGUMENT

> ORIGIN
> **Chapter III**
> *'The Struggle for Existence'*

We have already glimpsed a moment (see page 6) when the 'struggle' comes into focus in Chapter III of the *Origin*:

QUOTATION

What a struggle between the several kinds of trees must have gone on ... what war between insect and insect – between insects, snails and other animals.

cross reference P: 132; O: 62; ML: 102

---

and we have seen something of the influences which Darwin was absorbing at this point in his theory, notably from the ideas of Thomas Malthus. The 'struggle' has been placed in terms of the overall development of his great book. But there is a unique quality to Darwin's third chapter, entitled 'The Struggle for Existence'. This section of the *Origin* is distinguished by a particularly effective and heightened use of **persuasive logic** and **forceful argumentation**. The aims of the present account here are:

* to identify the key stages of Darwin's argument for 'struggle';

* to show how he makes such a persuasive case at the heart of the book;

* to indicate why 'the struggle for existence' is also one of the most influential passages anywhere in modern thought.

In a modern edition, the chapter on 'the struggle for existence' is fewer than twenty pages long, yet its images, ideas and phrases still reverberate in our culture, a century and a half after their first publication.

Darwin's question

To open his key chapter, Darwin again looks out over what he calls '*organic beings in a state of nature*', in other words the different forms of life which populate the earth, and which have not come into existence through human manipulation or for human ends. These 'beings', of course, include ourselves as well as '*the beetle which dives through the water*' and '*the plumed seed which is wafted by the gentlest breeze*'. The writing becomes lyrical as the gaze closes in on details of the living world:

> ## QUOTATION
> *We see these beautiful co-adaptations most plainly in the woodpecker and the misseltoe; and only a little less plainly in the humble parasite which clings.*
>
> cross reference P: 114; O: 51; ML: 88

As always, this gaze exerts pressure on familiar words: 'beautiful' takes on new shades of meaning when you apply it to the way a parasitic plant fits its niche in the world, or a bird is adjusted to its way of life. But the sense of beauty is not an end in itself. Through his way of looking, Darwin wants to define a question. He is interested 'in understanding **how** species arise in nature'. Darwin's gaze is not a way of accepting the world, but a way of questioning it.

In one of the great sentences from Darwin's book, we can actually watch the vision of nature turning into a scientific question:

CLOSE-UP

How have all those exquisite adaptations of one part of the organisation to another part, and to the conditions of life, and of one distinct organic being to another being, been perfected?

cross reference P: 114; O: 50; ML: 87

Anatomy of the close-up

o This sentence is at the heart of the *Origin*. Each word counts, not only in the book, but in the history of thought. In many ways, the first word is the most important: Darwin develops his key arguments as a response to a question, and the question is 'how', rather than 'why'. We are about to encounter an explanation of organic beings, but it isn't the kind of explanation you get if you ask the question 'why?'. Darwin, in fact, has no answers to the question of why the beetle and the seed have come into the world, or indeed why the world itself is there. To many people, the word 'origin' sounds like a 'why?' word: 'What does it all mean?', 'Where does it all come from?' To Darwin, 'origin' is a 'how?' word: his explanation is about the *process* which has shaped life forms, and not about the reasons for life itself.

What Darwin is interested in explaining are '**adaptations**', the ways in which things fit together in their specific contexts. As we have already seen, 'adaptation' has become one of the most important terms in what we think of as a Darwinian view. He is indeed asking how the parts of a beetle are so well adapted to one another, how the creature is so well adapted to its world. To that extent, adaptation is a technical word, part of the scientific theory. Yet 'exquisite' is a more religious word: the adaptations are beautiful as well as intricate, appealing as well as puzzling. It's as if a new language is growing alongside older languages, a new way of thinking alongside older thoughts and feelings.

One can see how Darwin is recreating this word by comparison with Lyell:

> *For in the first place, where a capacity is given to*
> *individuals to adapt themselves to new circumstances.*
>
> Lyell, op. cit., p. 211

Before Darwin, 'adaptation' is a capacity given by the Creator. After Darwin, adaptation belongs to a natural process. Lyell believes he knows *why* adaptation exists; Darwin thinks he knows *how* it arises.

The question, then, is *how* a seed comes to be so well suited to its function, of spreading the possible offspring of a plant as widely as possible. This seems like a technical kind of problem, which might give rise to a scientific theory. But in the end Darwin is still using words like '**perfected**' (a term which some may feel belongs to the religious vision of nature) to describe what he wants to explain in this way. For Darwin, if things have been perfected it is by an impersonal process, one which is still continuing. There is nothing fixed or final about his perfected beetle and seed: they are still changing, as the world around them changes.

The design of the 'struggle' argument

'The struggle for existence' is now in action: we have had a vision of nature and then a question to be answered. The following diagram shows how the argument will unfold through its crucial phases:

The 'struggle' argument

Presents a picture of nature :
(i) Diversity of 'organic beings'
(ii) Adaptation of each to its function

↓

Asks a question about that picture of nature:
How have adaptations emerged?
Defines a key term for giving an explanation
adaptation and diversity arise from '*struggle* for existence'

↓

Relates the key term to Malthus's established theory
Struggle is due to the rate at which all beings reproduce

These four steps take only a few pages, yet they are the leaps by which Darwin launches the first phase of his great explanation.

First, then, Darwin has presented **pictures of nature**, an earth populated by 'exquisite adaptations'. But in the same moment, he shows himself wondering **how** these beings happened. Then there is a twist, and the argument moves on to another phase, the hypothetical **explanation**. Darwin insists that the key to the puzzle is what he first calls 'severe competition' (**P**: 115; **O**: 52; **ML**: 89) and then labels this with the term 'struggle for existence'. Here, as the map shows, he pauses to **define the terms** of the argument. This is one of Darwin's great strengths both as a scientific theorist and as a persuader: he keeps looking at the words he is using, asking what they mean for us, attuning the old language to his new way of looking at the world.

In this definition, Darwin admits that he is using his key terms '*in a large and metaphorical sense*'. In his argument, 'struggle' will have many different meanings. First, and most oddly, 'struggle' will include what he calls '*dependence of one being on another*'. Yes, he agrees, the most obvious kind of struggle is that between two canine animals competing for the same food. But all struggle is the other side of being dependent. So a plant in a dry environment may struggle against the ferocious climate, but only because it is also dependent upon water. Those fighting dogs are struggling because each depends upon the same food sources, which have become scarce.

This is one of the most important moments in Darwin's argument, this redefinition of the meaning of struggle. For Darwin, 'struggle' is a way of talking about the interdependence of all life forms, and about the active relationships between them. True, he has deliberately chosen a bleak word, but he does not really use it in a depressing way. Struggle is about activity and change, about the possibility of growth, as well as about limitation and threat. The counterpart of 'struggle' is always dependence: so Darwin says that mistletoe plants on a single tree are struggling with one another because each depends upon the same tree. For Darwin, trees are often symbols as well as examples. Here the host tree, with the parasites on it, seems to stand for the whole environment of which it is part. Struggle is the sign that organisms share a finite world, not that they are inherently aggressive or driven by a will to destroy one another.

One other aspect of struggle is that it includes successful reproduction as well as individual survival. Reproduction leads on to the next step in the argument, where Darwin makes his link with an established theory, namely the ideas of Thomas Malthus who was introduced as an influence in our survey of Darwin's work. Having begun to create his own language, with his definition of struggle, Darwin then turns back to the terminology of others. He claims that struggle used in his way, may be a good starting-point to explain the forms of life on earth. He will shortly give more elaborate evidence. But first he wants to suggest some reasons why the struggle occurs, and he takes these reasons from his predecessor Malthus in a crisp sentence:

QUOTATION

A struggle for existence follows from the high rate at which all organic beings tend to increase.

cross reference P: 116; O: 53; ML: 90

Every life form is driven to expand its numbers. If each individual were successful, the earth would be overrun by the offspring of a single species, or even a few individuals. Darwin calls this principle, 'the doctrine of Malthus', but 'applied to the whole animal and vegetable kingdoms.' As we saw, Malthus was interested in human population growth, which he argued was geometric whereas food supplies would only grow in an arithmetic progression. In other words, numbers would multiply, whereas supplies would only add up: three times three as against three plus three, five times five as against five plus five. The gap widens, and the excess population must perish. Malthus drew all kinds of moral conclusions from this, but Darwin showed no such interest. He simply uses the doctrine as a way of suggesting why struggle is necessary, not in a moral sense, but physically. Without struggle, 'the earth would soon be covered' by the offspring of 'a single pair', whether of elephants, ants or daffodils.

In a few paragraphs, we have passed from the exquisite forms of life, through the universal struggle to the law of population increase which enforces that struggle. The whole argument is that the struggle *explains* the countless adaptations, and the population pressure explains the struggle. It is in this competitive context that a difference between individuals can provide an advantage, both to that individual and, in time, to the species. Darwin launches this aspect of his theory with beautifully economical logic, and he also combines his own original proposal with one of the most established and influential ideas of the time.

Darwin the persuader

This section of Chapter III shows Darwin at the height of his argumentative powers. Everything that we have seen so far of the formation of the Darwinian gaze, and the refinement of the Darwinian language, goes into shaping these moments of critical argumentation. Here are three ways to appreciate the significance of these arguments:

1 The arguing itself is a work of both art and science, a structure at once beautiful and persuasive. One can still appreciate this writing as a living process of **persuasive communication**.

2 The vision spreads out into almost everything that has been thought about both nature and society in succeeding generations. Some have agreed, others have passionately dissented, and everyone has had their own interpretation, but you can feel here that you are approaching **the source** of philosophy and politics, ethics and storytelling in modern times.

3 Almost every phrase in 'The Struggle for Existence' conveys something distinctively of its own time. You can also read these pages as condensed expressions of **the spirit of their own age**.

The panorama of struggle

'Struggle for existence' has now been put in place and we have the first half of the great explanation. Before he moves on to add the second part, the concept of natural selection, Darwin makes another characteristic move. He uses the concept of struggle for existence to create a panoramic vision of life on earth. The gaze has shaped an argument, and now the argument feeds back into the gaze.

The panorama begins with a view of the whole earth:

> QUOTATION
>
> *Lighten any check, mitigate the destruction ever so little, and the number of the species will instantaneously increase to almost any amount. The face of Nature may be compared to a yielding surface, with ten thousand sharp wedges packed close together and driven inwards by incessant blows, sometimes one wedge being struck, and then another, with great force.*
>
> cross reference P: 119; O: 56 omits 'face of nature'; ML: 94 likewise

Those 'blows' are the population pressures of each species as it adapts and expands; the 'face of nature' stands for the earth itself, and for the system of organic life as a whole.

Darwin pauses to think about this vision. We now see the whole, but we should not presume that we have truly understood even the smallest of the details:

> QUOTATION
>
> *We know not exactly what the checks are in even one instance ... even in regard to mankind.*
>
> cross reference P: 120; O: 56; ML: 94

In the work of Malthus, the 'checks' on human population are confidently, even arrogantly, specified as misery, vice and disease. Darwin extracts the Malthusian word checks and gives it a different quality. In the Darwinian language, a check is a mysterious thing, elusive and subtle. It can't be imagined like a great barrier; rather, it is a web of indefinable connections, invisible pressures that halt the growth of every single species on its path to infinite expansion. Darwin's gaze *deepens* the world, gives it a shadowy and subtle feeling. From this moment of mystery, the gaze expands again to give a confident panorama across space:

QUOTATION

When we travel from south to north, or from a damp region to a dry, we invariably see some species getting rarer and rarer, and finally disappearing,; and the change of climate being conspicuous, we are tempted to attribute the whole effect to its direct action. But this is a very false view; ... constantly suffering enormous destruction .... from enemies and competitors for the same place and food; and if these enemies and competitors be in the least degree favoured by any slight change of climate, they will increase in numbers.

cross reference P: 121; O: 58; ML: 96

We begin with the Darwinian moment of sight, 'we invariably see', and then there is the argumentative move: 'we are tempted to attribute' what we see to a simple cause, like climate, but this is 'a very false view'.

It is not the changed climate of the northward journey that in itself reduces, say, the number of a more southerly tree. It is the way other life forms respond to that climate change, making inroads into the available resources of the limited earth. In this passage, the word 'view' has a significant double sense. It means a judgement, but it also carries some of its more primary meaning, something you see, a panorama. In this word view, we have both an explanation and a gaze. The false 'view' is the product both of not seeing and of not thinking. In the Darwinian look, you cannot separate these two powers: you achieve a new focus on the world in a moment of simultaneous sight and thought.

As the Darwinian gaze passes across the planet, following the struggle from place to place, it pauses to pick out individual episodes. At one point, for instance, we are in Paraguay: 'here neither cattle nor dogs nor horses have ever run wild'. In the Darwinian panorama of struggle, you also need to see what is *not* there. Once you are used to seeing the absences, such as the lack of the Paraguayan wild cow, then that incites the explanation:

QUOTATION

...this is caused by the greater number in Paraguay of a certain fly, which lays its eggs in the navels of these animals when first born.

cross reference P: 124; O: 61; ML: 100

So the gaze, passing across 'the face of Nature', or roving 'from south to north', pauses and lights on this 'certain fly' which is laying eggs in the navels of Paraguayan cows. If there are fewer flies, there will be more cows, but this doesn't seem to happen. Why not? Darwin theorizes that, '*the increase in these flies ... [is] checked ... probably by birds*'. So more insects means fewer flies, which would mean more cattle, some running wild. Then you can ask whether anything could change the number of birds. This sort of chain of connections typifies a Darwinian explanation: it is not a single, unitary conflict, but an intricate or entangled mesh: '*Battle within battle.*'

At this point in the panorama, we have the passage which we examined earlier about the entangled bank, the Indian mound and the virgin forest, where Darwin sees in '*the ancient Indian mound ... the same beautiful diversity and proportion of kinds as in the surrounding virgin forest.*'

In context of the whole argument, one can see that these landscapes are another embodiment of the whole vision of the earth. That 'entangled bank' is Mr Darwin's planet, and the Indian mound seems to stand for something important about the place of humanity on that earth. Given that Darwin restrains himself from saying much directly about humanity in this book, that image seems quite pointed. The virgin forest reappears, it reabsorbs the human creation. The struggle moves over the remains of a human culture.

EXPLANATION 2: THE NARRATIVE OF NATURAL SELECTION

Darwin devotes his fourth chapter to presenting the central concept of natural selection, refining and defending what has so far been achieved. We will now map Darwin's varying definition of natural selection and examine its role in strengthening his explanation of how species develop and change through struggle.

> O R I G I N
> **Chapter 1V**
> *'Natural Selection'*
> (The later editions, as in **ML**, add 'or the survival of the fittest')

A Darwinian narrative: climate change and border controls

The concept of natural selection adds a new dimension to the Darwinian way of looking. It is this concept which turns the scenes into stories, and makes the Darwinian panorama we have witnessed into a narrative history of life on earth. If the third chapter is the moment of argument, the fourth chapter of the *Origin* is the moment of narrative.

Darwin sets up his account of natural selection by imagining a place:

> QUOTATION
> *...taking the case of a country undergoing some physical change, for instance, of climate.*
>
> cross reference P: 131; O: 68; ML: 109

This place is being altered, a new environment is emerging:

QUOTATION

The **proportional numbers** *of its inhabitants would almost immediately undergo a change, and some species might become extinct.* [Author's emphasis]

cross reference P: 131; O: 68; ML: 104

However, the story Darwin tells is not about the way climate affects living beings: it is about **the way life forms affect each other**, in the context of changing situations. This Darwinian story is a close relative, as the critic and historian Gillian Beer has shown, of Victorian novels. Emily Brontë's *Wuthering Heights*, for example, has a place in its title and sets every event in the context of the moods and impacts of that world. But the story itself concerns the way the characters affect each other as they share – or try not to share – this space. Again George Eliot's *Middlemarch* shows how people impact upon one another, in the context of their changing physical and social environment.

You cannot say, in these narratives, that the setting simply causes the action. The logic is more indirect. The changing environment impacts upon the unfolding relationships between the characters. The same logic applies in Darwin's narratives of natural selection: '… *any change in the numerical proportions of some of the inhabitants, independently of the change of climate itself, would most seriously affect many of the others.*' Climate affects each inhabitant via its impact on all the others.

In many Victorian novels, the action begins with the appearance of a stranger in a hitherto closed setting. This is true in *Wuthering Heights*, in *Oliver Twist* and many Dickens novels, in *Middlemarch* and in many of Hardy's stories. Darwin, too, is interested in the impact of intruders on systems as they begin to change: '*If the country were open on its borders, new forms would certainly immigrate.'*

He is also, like the novelists, interested in relatively closed systems, where there are few intrusions, and where most of the changes occur internally:

QUOTATION

But in the case of an island, or of a country partly surrounded by barriers, into which new and better adapted forms could not freely enter, we should then have places in the economy of nature which would assuredly be better filled up, if some of the original inhabitants were in some manner modified.

cross reference **P**: 131; **O**: 68; **ML**: 109

The key word here is **'modified'**. Just as Victorian novelists are interested in exploring the ways a character changes under the impact of altered relationships, so Darwin tells his stories in order to pinpoint the factors leading to life forms becoming 'in some manner modified'. Like the novelists, he is intrigued by the scope for change within 'the original inhabitants' of his chosen spot.

Victorian novels tend to tell stories with an element of **irony** about them. Often, what makes a story ironic is that something which seems trivial turns out to be the major turning point in the whole saga. The same is true in Darwinian narratives. It is the small detail which determines the outcome:

> QUOTATION
>
> *every slight modification, which in the course of ages* **chanced** *to arise, and which in any way favoured the individuals of any of the species, by better adapting them to the altered conditions, would tend to be preserved; and* **natural selection** *would thus have free scope for the work of* **improvement.** [Author's emphases]
>
> cross reference P: 131; O: 68; ML: 110 deletes 'chanced to'

A further aspect which Darwin's irony shares with his novelist contemporaries is that there is this scope for **chance** in the midst of systems that seem in many other ways tightly logical. It is enough for a tiny flicker of chance to intervene, and the balance of the whole narrative is changed forever.

As one can also see from this passage, Darwin also shares with the novelists a taste for ironies that deliver some degree of **improvement.** By the end of a Victorian story, some of the original inhabitants will have died or been defeated; others will have grown, perhaps in small ways or maybe more significantly. Details of character which seem trivial at the start will turn out to dictate the fate of characters, and thus to redefine the system of relationships within which those lives are lived. Natural selection, however, takes the place in Darwin's scientific narrative of more moralistic or mystical forces that might be seen to have presided over the outcomes of a Victorian novel.

Darwin's gaze, like Emily Brontë's or George Eliot's, tries to take into account 'every slight modification' which appears in the course of the changing relationships on his imaginary island. Natural selection is the concept which makes these slight differences relevant. We have a series of vivid passages where the way of seeing turns into the way of narrating, where the thinking observer becomes a master storyteller:

QUOTATION

For as all the inhabitants of each country are struggling together with nicely balanced forces, extremely slight modifications in the structure or habits of one inhabitant would often give it an advantage.

cross reference P: 132; O: 69; ML: 111

Here Darwin's gaze merges with natural selection's scrutiny, in picking out these 'extremely slight modifications', and making them into the critical facts for the story. We can also now see clearly how the two halves of the Darwinian explanation come together. It is the struggle which creates the relationships within which natural selection can pick out these apparently minor differences and work on them.

Darwin's narratives of natural selection

* The story is about **indirect effects**, of climate and physical conditions.

* Circumstances change **relationships**, which are the medium of all development.

* Strangers and migrants transform the whole system.

* These are tales of **modification** among the established inhabitants.

* Science recognizes that there is scope for **chance**.

* Darwin sees the **irony** of the small difference that transforms the wider outcome.

* Natural selection tells stories of **improvement**, though not as a whole grand narrative of progress.

The meaning of Darwin's narratives: the thickness of egg shells

The great passage on natural selection's gaze comes immediately after this narrative of the island: '...*natural selection is daily and hourly scrutinising ... every variation ... silently and insensibly working ... at the improvement.*'

In context, this definition is the finale to a great story. From that point, the *Origin* has its explanation in place and, as we will see, the problem then is to illustrate that theory and to acknowledge its current limitations. What Darwin does *not* offer, as he repeatedly emphasizes, is a single story of life's progress, as can be seen from an extract from Chapter X:

QUOTATION

I believe in no fixed law of development, causing all the inhabitants of a country to change abruptly, or simultaneously, or to an equal degree.

cross reference P: 318; O: 253; ML: 446 (minor variation)

Science belongs to its time, and expresses its discoveries within a cultural world. It is in that sense that Darwinian natural selection shares a cultural space with Dickens and George Eliot, just as Newtonian physics shared a space with, among others, the earlier novels of writers like Defoe, the philosophy of Hobbes and the poems of Dryden. In some arguments, there has been confusion about such connections. While it is true that scientific ideas are created with the materials of their age, it does not follow that science is simply 'a cultural construct', which merely imposes a view on the world which it claims to observe or explain.

Darwin's theory of natural selection does have immense explanatory and observational power, and part of that power derives from the richness of the cultural moment within which it emerged. It is not a coincidence that Darwin's narrative was conceived in a moment shared by some of the most complex realist stories in the history of world literature. After all, novels, too, can have their truth-telling powers. Like his fiction-writing contemporaries, Darwin recognized the ambiguity of stories. He too acknowledged the importance of perspective. For example, how does it come about that baby birds can crack their shells so easily? Natural selection is at work, yes, but there are two equal ways to tell the story:

QUOTATION

the hard tip of the beak of nestling birds ... there would be simultaneously, the most rigorous selection of the young birds within the egg, which had the most powerful and hardest beaks ... or, more delicate and more easily broken shells might be selected.

cross reference P: 135; O: 72; ML: 116

The Darwinian gaze comes to rest on 'the hard tip' of those tiny beaks. Have they been adapted to the eggshells? Or is it the other way around? Every organic entity tells a story, or rather several possible stories. Natural selection is the cue for this vast empire of storytelling, not the short cut to a single grand tale.

Sexual selection

The last part of the account of natural selection reinforces the link with Victorian fiction. Darwin next considers

---

CLOSE-UP

what I call Sexual Selection. This depends, not on a struggle for existence, but on a struggle between the males for the possession of the females.

cross reference P: 136; O: 73; ML: 117

Some characteristics result from the struggle 'between the males' within a species. Alternatively, Darwin recognizes, one could see the same characteristics as resulting from the likes and dislikes of the females:

QUOTATION

Those who have closely attended to birds in confinement will know that they often take individual preferences and dislikes.

cross reference P: 137; O: 74; ML: 119

The key phrase here is: '*those who have closely attended*'. The scientist has to try to tell the story as if from that female blackbird's point of view – and only those who have watched carefully will have noticed the existence of that perspective. In fact, Darwin adds, when we appreciate the song of the blackbirds, we are actually entering into the point of view of countless generations of female birds in, '*... selecting, during thousands of generations, the most melodious or beautiful males*' (cross reference **P**: 137; **O**: 74; **ML**: 119).

Modern genetics has taken this theory of sexual selection and applied it to the microscopic domain, where, as Steve Jones says in *Almost Like A Whale* (p. 83): '*Many of Nature's most attractive features – flowers, birdsong, mandrill bottoms – result from rivalry among male sex cells for access to eggs.*'

Darwinian Gaps 5

DARWIN AS SCIENTIFIC PROPHET

Chapters 1–IV of the *Origin* present an extraordinary number of positive propositions about the natural world. Darwin defines, illustrates and defends his new theory; he launches the key arguments and tells the new stories. The double explanation – 'struggle for existence' and 'natural selection' – is now in place. Chapter V onwards play a different role. Virtually all of the second half of the *Origin* is concerned with exploring what can be called 'the Darwinian gaps'. In a whole variety of ways, Darwin realized that his account was incomplete. This is the other extraordinary dimension of the Darwinian gaze – this ability to see the gaps in his own theory, or to recognize where the world remained unaccounted for.

There is a paradox to the achievement of the *Origin*. It is in these more negative chapters that Darwin comes closest to being a scientific prophet. For he has a basic confidence that the gaps in his theory can be filled, in time. Where he identifies the missing links, he often uses quite a negative way of talking: we do not know, we cannot say, we cannot explain … Yet there is a fundamental confidence underlying these apparently negative admissions: there will come a time when the links are supplied. There is a whole deeply felt approach to knowledge, and particularly to the theoretical aspect of scientific knowledge, implied by the way Darwin identifies and exposes the missing elements in his own account of the natural world.

The aims of this chapter are:

* to survey Darwin's Chapters V onwards;

* to show how in this part of the *Origin*, Darwin provides various accounts for the gaps in his theory;

* to explain Darwin's wider approach of scientific knowledge;

* to convey Darwin's overview of life on earth, 'Mr Darwin's Planet';

* to explain the continuing scientific influence of Darwin's ideas.

GAP 1: THE MISSING LAWS

ORIGIN
Chapter V
'Laws of Variation'

If an ordinary book announced as a heading the 'Laws of Variation', one might expect it to proclaim those laws. Remember how crucial variation is to Darwin's whole way of seeing the world: it is the term which covers precisely those differences on which natural selection can work. No variation would mean no selection. Yet Darwin makes clear that the laws of variation are unknown to him, and to his contemporaries. His point is rather that *there must be* such laws. He is creating an agenda for the evolution of his own theory and explaining what the missing elements must be like, given the nature of what is already known.

A lesser genius would probably have improvised some laws of variation at this point. Darwin had such incredible facility with concepts and explanations that he could certainly have devised a plausible and persuasive scheme. In fact, he refuses to improvise: Darwin's genius includes his awareness of when *not* to have an idea.

Darwinian Principles of Knowledge 1
Theory must not become an excuse to stop looking. You cannot substitute a general explanation for specific information.

For Darwin, general theory is never a substitute for specific knowledge. Even the best theory in the world does not replace the need to gather new information. On the contrary: the theory serves as a guide to further research. Darwin gives an example of how his theory reveals new questions.

Winter fur

If you look at an animal in a certain environment, you may see a feature which seems advantageous. However, you cannot instantly assume that you are in a position to explain this feature, even on the basis of the authoritative concept of natural selection.

QUOTATION

When a variation is of the slightest use to a being, we cannot tell how much of it to attribute to the accumulative action of natural selection, and how much to the conditions of life.

cross reference P: 175; O: 110; ML: 174

The advantage *may* be due to natural selection and may have been inherited from similarly favoured parents. But useful characteristics can also arise as direct responses to the circumstances themselves, as an immediate physical response. Such cases lie outside the range of natural selection, which only works where an inherited quality turns out to be an advantage, and as such is more likely to be passed on. Darwin has absolutely set aside, as we have seen, any notion that a characteristic acquired through life may be passed on. Only an inherited characteristic can itself be passed on.

Take a species of furry mammal, with some individuals living in a cold climate, others in more temperate places. As ever, Darwin pays homage to the non-scientific expertise of the professionals: '*It is well known to furriers that animals of the same species have thicker and better fur the more severe the climate is under which they have lived.*'

Dogs in a colder place have woollier coats. Isn't this a good example of a variation which fits with the theory of natural selection? Not necessarily, says Darwin: '... *but who can tell how much of this difference is due to the warmest-clad individuals having been favoured and preserved during many generations, and how much to the direct action of the severe climate?*' Maybe there is something about the biology of fur-growing which tells a wolf to get furrier when the climate is colder. In that case, all the individuals may possess that fur-cladding thermostat, and it is operating equally in those who have *not* grown more fur as in those who have.

Then there is an even more radical question for Darwin. Why do these small differences like thicker fur arise in the first place? If you look at the offspring, they may well display characteristics different from the parents – one little wolf may be woollier than the others. We may then be observing a variation which could be inheritable by *their* offspring. But Darwin also knows he has no way of explaining when or how such mutations arise:

QUOTATION

Our ignorance of the laws of variation is profound. Not in one case out of a hundred can we pretend to assign any reason why this or that part differs, more or less, from the same part in the parents.

cross reference P: 202; O: 137; ML: 209

This is one of the Darwinian gaps in which modern genetics has grown: the missing laws of variation. Modern science has been able to build up, as Steve Jones shows in his updating of the second part of Darwin's book, many of the additional explanations.

It is significant that when Jones wrote his wonderful modernization of the *Origin*: (*Almost Like A Whale: The Origin of Species Updated*), he gave far more space to the second half of the book. This is where

Darwin identifies the missing elements in his theory so far, and it is a sign of how accurately Darwin saw the missing links, that a contemporary authority can in many cases fill them in. The dialogue between Jones and Darwin is moving because it so often has its origins in cues given by the Victorian, cues like recognizing that he can give no '*reason why this or that part differs*'.

Of course, Darwin does sometimes speculate in these areas beyond his means and Jones (p. 111) shows why the limitations of his conclusions have not damaged the standing of the *Origin* as a whole. In the case of heredity, Darwin's gaps are still being filled in. As we have seen, the crucial development has been the concept of 'the gene', and the vast field of research that it has opened up. This field only began in the later nineteenth century, with the work of Gregor Mendel, but by recognizing the gaps, Darwin had already prepared his theory to travel into that future.

GAP 2: THE MISSING LINKS

One kind of Darwinian gap arises where the theory itself has missing conceptual elements, like the **gene**, which of course also correspond to gaps in knowledge. The missing laws of variation are an example.

> O R I G I N
> **Chapter VI**
> '*Difficulties on Theory*'

Another type of Darwinian gap is due to the inaccessibility of the data itself. For example, the story of natural selection suggests to Darwin that there must have been transitional stages by which any given characteristic developed. Say there is a pigeon with an elongated eyelid; you would expect to find previous examples of slightly elongated eyelids, and then of longer ones and so on, as natural selection, or human selection in that case, adds up the differences.

QUOTATION

... as by this theory innumerable transitional forms must have existed, why do we not find them embedded in countless numbers in the crust of the earth?

cross reference P: 206; O: 141; ML: 213

Where are these intermediate stages – or their fossil remains? Again, Darwin knows his limits:

Darwinian Principles of Knowledge 2

A perfect theory would require perfect knowledge. In the absence of perfect knowledge, theory is a way of recognising the limits of present knowledge.

In many cases, you would expect to find transitional forms, say a fossilized pre-elephant with a slightly longer trunk, or teeth more like tusks, and often Darwin could not find them. But that does not disprove the theory: '*the answer mainly lies in the [geological] record being incomparably less perfect than is generally supposed.*'

One of Darwin's great strengths is that he is as determined to explain these gaps as he is to fill in the positives. If the transitional forms are often hard to find, there may be an additional reason, beyond the imperfections of geology:

QUOTATION

the more common forms, in the race for life, will tend to beat and supplant the less common forms, for these will be more slowly modified and improved.

cross reference P: 210; O: 144; ML: 217

The transitional forms prove hard to find, because they were so completely replaced. Even their fossils are comparatively rare. Darwin again moves to an example, this time of an imperfect sequence.

The eyes of crustaceans

In a theory so focused on looking and seeing, the eye is bound to be particularly interesting. Surely, Darwin begins, it is hard to imagine the human eye gradually emerging through slow modifications of previous forms of eye? Isn't this organ simply too wonderful to have been evolved in such an apparently haphazard way?

QUOTATION

To suppose that the eye, with all its inimitable contrivances for adjusting the focus to different distances, for admitting different amounts of light ... could have been formed by natural selection, seems, I freely confess, absurd in the highest possible degree.

cross reference P: 217; O: 152; ML: 227

Emma Darwin may be the direct source of this anxiety. Desmond and Moore paint a picture of her sitting with a pile of Darwin's manuscript pages, pencil poised. She noted unclear passages and raised questions. One of her most forceful doubts was pencilled in when she reached a tentative statement about the eye: '"*A great assumption/E.D.*" *she scribbled against his claim that the human eye "may possibly have been acquired by gradual selection of slight but in each case useful deviations"*' (Desmond and Moore, p. 19).

However, if the theory of natural selection works, then it must work for eyes as well as every other organ or organism. We must be able to find evidence that the form of the eye has been modified as advantageous variations take hold. Darwin looks back into deeper time and his gaze comes to rest on a host of tiny developments:

QUOTATION

In certain crustaceans, for instance, there is a double cornea, the inner one divided into facets, within each of which there is a lens shaped swelling. In other crustaceans the transparent cones which are coated by pigment ... are convex at their upper end ... these facts ... show that there is much graduated diversity in the eyes of living crustaceans.

cross reference P: 218; O: 153; ML: uses other example here

In other words, even the eye has been capable of gradual transition from less effective to slightly more effective forms. It is because Darwin has such a strong idea that there must have been a sequence, that he realizes the links are missing. He cannot actually find a whole series of gradually improving eyes, but the sequences which can be identified, sequences which Darwin calls '**graduated diversity**', are enough to show what must be missing, and what to look for in the future.

Following on, a century and a half later, Jones (pp. 139–40) is then able to fill in many such missing links. He points to the eyes of night-flying insects that '*have large lenses that increase sensitivity by a hundred times*', of bees which '*have an upright strip of sensors adapted to the vertical world of trees*', and of water-skiers whose '*eyes are adapted to the flatness of their world*'. Subsequent data has shown that the eye has evolved with natural selection's usual inventive messiness, as Darwin foresaw.

Darwin concludes his account of the eye by *closing* a gap:

> QUOTATION
>
> *If it could be demonstrated that any complex organ existed, which could not possibly have been formed by numerous, successive, slight modifications, my theory would absolutely break down. But I can find no such case.*
>
> cross reference P: 219; O: 154; ML: 232

The full facts have not yet emerged; but no facts have appeared to refute the theory.

GAP 3: HOSTILE GAPS

The next chapter of the *Origin* concerns what can be called 'a hostile gap', that is one which, if it turned out to be unfillable, would be enough to subvert the whole explanation.

> ORIGIN
> ## Chapter VII
> *'Instincts'*
> ML Ch VIII – after an added chapter on further objections)

> Darwinian Principles of Knowledge 3
> *A universal theory cannot afford any counter examples, though it can afford endless missing links.*

Now Darwin faces the question of 'instincts', an issue that remains controversial in the twenty-first century. He begins by distinguishing the loose sense of something being '**instinctive**' from the true meaning. People often talk of someone having an instinctive ability, say at music:

> QUOTATION
>
> *If Mozart, instead of playing the pianoforte at three years old with wonderfully little practice, had played a tune with no practice at all, he might truly be said to have done so instinctively.*
>
> cross reference P: 235; O: 170; ML: 319

A true instinct is a behaviour that is not in any sense acquired though learning or experience, though of course it then interacts with both knowledge and experience. The question is: can natural selection, in the context of the struggle for existence, account for instincts? Insofar as instinctive behaviour is inherited, then it does fall into an area which matters for the theory. If there is no way to account for instincts in terms of natural selection, that would be what we can call a 'hostile gap' in the theory, and it would call for a revision of the whole way of looking that has been so carefully built upon.

If instincts do arise through natural selection, then they emerge gradually, through the slow modification of previous patterns:

> QUOTATION
>
> *No complex instinct can possibly be produced through natural selection, except by the slow and gradual accumulation of numerous, slight, yet profitable, variations.*
>
> cross reference P: 236; O: 171; ML: 320

Those words, '*slight, yet profitable, variations*' take us back to the Darwinian gaze. There can be, from this point of view, no division between instincts and other inheritable variations.

The wonderful instinct of making slaves

Darwin zooms in on some ants. He observes two species which he calls 'slave-making ants'. They are in different places, and they behave very differently. Each keeps in its nest a group of alien ants, which perform many of the basic labours of the ant society: hence the term, 'slave-making'. Such behaviour is a perfect example of what Darwin means by an instinct. These types of ant keep their slaves entirely as an inherited pattern. But the activity involved seems so rich and complex: can natural selection truly account for these ant worlds?

Darwin argues that the reason he can work on these cases at all is that there is more than one slave-maker:

QUOTATION

If we had not known of any other slave-making ant, it would have been hopeless to have speculated how so wonderful an instinct could have been perfected.

cross reference P: 244; O: 179; ML: 334

One type of ant is *Formica rufescens*: '*So utterly helpless are the masters ... without a slave ... they could not even feed themselves.*' But the other is far more independent of its slaves, and is called, forbiddingly, *Formica sanguinea* (*sanguis* is Latin for blood). In this second case, '*the slaves are black and not above half the size of their red masters*' and '*the masters ... may be constantly seen bringing in materials for the nest, and food of all kinds*'. Darwin needs the comparison because this kind of data enables him to think in terms of 'variations'. Once there are these alternative forms of slave-maker, he can begin to look at each as a different variant from a lost original.

Throughout his life, Darwin was a passionate opponent of human slavery. It was the political and moral issue on which he was consistently most outspoken. For example, when he was a young scientific officer of *The Beagle*, he risked alienating the captain, no small matter in the circumstances, by expressing his rooted objections to slavery. It is hard not to hear some degree of quiet irony in Darwin's phrase about '*the wonderful instinct of making slaves*' that follows the description of these ants.

However, the deeper point is that Darwin does *not* seek to draw parallels with human society, unlike so many later discussions of animals' behaviour. He did, in subsequent works, write movingly about emotion in man and animals, and he wanted to see humanity in its place in the natural world. But he did not exploit instances like these ants to tell simple parables about humanity. This refusal to moralize the data recalls a passage from the Victorian radical John Stuart Mill:

> *A great multitude of people are continually talking of the Law of Nature; and then they go on giving you their sentiments about what is right and what is wrong: and these sentiments, you are to understand, are so many chapters and sections of the Law of Nature.*
>
> Mill, J.S. *On Bentham* (1838)

Darwin's 'wonderful instinct' is a good example of how he refuses to read his '*sentiments about what is right and what is wrong*' into the book of nature. The Darwinian gaze would be impossible if Darwin were anxious to draw moral conclusions from the natural world.

One of the reasons why Darwin's approach – in such contrast to most natural history programmes – does not lend itself to moralistic parables is that he is concerned with 'differences' and variations. Instead of making moral points about instinct and nature, Darwin has another purpose. He is pursuing the *origins* of these instincts, not looking for their moral. He begins with a disclaimer: '*By what steps the instinct of F. sanguinea originated I will not pretend to conjecture*' (cross reference **P**: 247; **O**: 182; **ML**: 338).

But then he sketches the likely nature of the answer, even though he knows the data is not yet available:

> QUOTATION
>
> *But as ants, which are not slave-makers, will, as I have seen, carry off pupae of other species, if scattered near their nests, it is possible that pupae originally stored as food might become developed; and the ants thus unintentionally reared would then follow their proper instincts and do what work they could.*
>
> cross reference P: 247; O: 192; ML: 338

This is another fine Darwinian passage. On the one hand, a gap opens out: '*I will not pretend to conjecture*', he says. We do not know the answer. Then again, he wants to show what kind of answer can be expected. Darwin knows what would be sufficient to refute his theory, and how far he must go to meet that challenge. He sketches the research path for an account of the origins of instincts in natural selection. The likely answer will be to do with a chance event, which worked its way into the existing pattern of behaviour of both the slavers and the slaves. Presumably the implication is that those ant colonies which reacted more effectively towards the strangers – for whatever reason – might gain an advantage. The tendency to behave in that way would be inherited, given their greater success.

Darwin and the Theory of Instincts

If one looks at the relationship between Darwin and his predecessors, especially the admired Lyell, one can see why the question of instinct required at least a hypothetical answer:

> *The same remarks may hold true of instincts; for if it be foreseen that one species will have to encounter a great variety of foes, it may be necessary to arm it with great cunning and circumspection.*

Lyell, op. cit., p. 202

Lyell was in many respects an extremely radical thinker. But when it came to instincts, he fell back to an approach that was highly orthodox. Instincts are given in advance, complete and final:

> *When such remarkable habits appear in races of this species [dog], we may reasonably conjecture that they were given with no other view than for the use of man and the preservation of the dog which thus obtains protection.*
>
> Lyell, op. cit., p. 214

For Lyell, instincts reveal the hand of the Creator, or at least His unifying plan. Darwin realized he could not afford to leave this space open for religious explanations.

Again, Darwin turns out to have been a skilled anticipator of future research. Later Darwinian versions of instinct resemble the original – even though Darwin knew he was operating at the edge of his expertise. Here is an extract from Richard Dawkins on the pit-digging behaviour of antlions:

> *Probably an ancestral antlion existed which did not dig a pit but simply lurked just beneath the sand surface … Later, behaviour leading to the creation of a shallow depression in the sand probably was favoured by natural selection … became deeper and deeper.*
>
> Dawkins, *The Extended Phenotype* (1982), p. 20

The story matches Darwin's hypothesis about the slave-makers. By chance, some antlions are born with a tendency to make dips in the sand. They catch more ants. The characteristic is passed on in greater numbers.

Dawkins then adds a layer of later concepts: '*My point is that none of that history, nor any comparable history, could possibly have been true unless there was genetic variation in the behaviour at every step of the way.*' He argues that if a particular antlion behaves in a certain way, it must have some genetic cue for doing so. Therefore, in Dawkins's view, the ultimate explanation of natural selection is at the genetic level. It is the gene for the behaviour that is favoured, rather than the animals who behave in that way. Not everyone in the Darwinian field accepts this added argument: Gould in particular resists the further step to the genetic level of natural selection and insists that it cannot be genes which are selected, but only individuals with certain characteristics. We will see more of this controversy in the next chapter of this guide. This is a good example of how Darwin's initial attempt to fill a hostile gap has in turn given rise to a whole field of later arguments, generating in the process many important ideas and much central research.

GAP 4: COMPLEXITY GAPS

On the look-out for problems, Darwin next confronts what he calls '**hybridism**', the cross-breeding between individuals from distinct species. Can infertility across species divisions be a barrier designed to preserve in

> ORIGIN
> ## Chapter VIII
> '*Hybridism*'
> (ML: IX)

their purity the original types? This is the question that Darwin anticipates. His counter-argument is that hybridism is far too complex to have been understood properly in the current state of knowledge. In fact, he asserts, fertility and infertility cannot define a species. Sometimes, individuals from distinct species can interbreed; at other times, individuals from the same species will be sterile together:

QUOTATION

the sterility of various species when crossed is so different in degree and graduates away so insensibly, and, on the other hand, ... the fertility of pure species is so easily affected by various circumstances, that for all practical purposes it is most difficult to say where perfect fertility ends and sterility begins.

cross reference **P**: 266; **O**: 201; **ML**: 363

Here again we find the language of Darwinian gaps: *'for all practical purposes, it is most difficult to say'*. This is an example of what can be called 'a complexity gap': a phenomenon appears to be so complex that it cannot fit within any of the current models of explanation, let alone be understood in detail. Darwin is not afraid of such gaps.

MR DARWIN'S PLANET

In the chapters that follow, Darwin gazes across the earth. What he sees might fairly be named 'Mr Darwin's Planet'. The key features of this planet are its depth in time, and its intricacy in space. How, he asks, does his theory measure up against this planet as a whole?

Darwin's deep time

O R I G I N
Chapter 1X
'On the Imperfections of the Geological Record' (ML: X)

Inspecting the earth, Darwin reflects on the extent of the missing links in the chain of life forms that is implied by the stories of natural selection:

QUOTATION

... geology assuredly does not reveal any such finely graduated organic chain; and this, perhaps, is the most obvious and gravest objection which can be urged against my theory.

cross reference P: 292; O: 227; ML: 406

Where are all the fossil signs of transitions? Where is the evidence for the steps between what have since become distinct species but were once varieties of a single parent species? Darwin's basic answer is that the sheer depth of time makes it inevitable that most of the swarming life of the past has left no continuing trace.

His gaze turns to the present world, in one of the most personal and at the same time scientific passages:

QUOTATION

It is good to wander along lines of sea-coast, when formed of moderately hard rocks, and mark the process of degradation.

cross reference P: 294; O: 229; ML: 409

The words 'it is good to wander' contain the most touchingly intimate glimpse of Darwin that we get throughout the whole of the book. When you wander by the sea, you *'will be most deeply impressed by the slowness with which rocky coasts are worn away'*. This slowness is the sign of just how long must be the timespan over which the beaches have been formed, the cliffs been worn away. No one who seriously understood this timescale could expect to find comprehensive evidence of natural history.

We began this guide with pictures of Darwin looking – here in the closing stages of the great work itself, we glimpse the author as a solitary figure in a vast landscape. Is it a pure coincidence that here the words 'it is good' reappear: the same words which the authorized version of Genesis uses to articulate the gaze of the deity? If there is a spiritual dimension to the Darwinian sensibility, it is perhaps in the sheer sense of the scale of time, and the magnitude of what has passed away without trace on the earth.

Darwin then turns to the geological layers of the planet. These slices are by no means continuous: each represents a distinct chunk of massive earth-time, and there are missing ages which have not left behind such distinct traces. These are not the rings of a tree: in some phases, the process of degradation was more complete, in others the rock was more recalcitrant. But, Darwin says, it takes far longer for a species to mutate than is represented by one such slice of earth history:

> QUOTATION
>
> *Although each formation may mark a very long lapse of years, each perhaps is short compared with the period requisite to turn one species into another.*
>
> cross reference P: 302; O: 237; ML: 422

The history of many species simply is lost: the links which have gone into their formation took so long to change and vary that they have left no trace in the geology.

Extinction: lost worlds

ORIGIN
Chapter X
'On the Geological Succession of Organic Beings' (ML: XI)

Now the book comes to one of the most controversial issues in the Victorian period: the question of extinction. Here we are dealing with the kind of gap we called 'missing laws'. Darwin argues that there is no possibility as things stand of devising a single law that consistently explains the facts of extinction. Species pass away because of complex factors involving a whole network of relationships with other living forms:

> QUOTATION
>
> *On the theory of natural selection the extinction of old forms and the production of new and improved forms are intimately connected together.*
>
> cross reference P: 321; O: 256; ML: 449

These effects are elusive and so: '*No fixed law seems to determine the length of time during which any single species … endures.*'

Again, there is a little glimpse of Darwin the man, finding a fossil tooth in South America. He remembers his surprise when it seems to belong to a horse, since all the fossils around are from more exotic animals, like the mastodon. He discovers that this is not a modern horse at all, but an extinct ancestor. Then he asks why a species should be rare, and then become extinct: '... *why* ... *we answer that something is unfavourable in its conditions of life; but what that something is, we can hardly ever tell.*'

No law or theory will give a short cut:

QUOTATION

It is, indeed, quite futile to look to changes of currents, climate, or other physical conditions, as the cause of these great mutations in the forms of life under the most different climates.

cross reference P: 327; O: 262; ML: 457

ORIGIN
Chapter XI
*'Geographical
Distribution'*;
Chapter XII
*'Geographical
Distribution'* (cont'd)
(ML: XII)

Darwin concentrates on ruling out alternatives to his own theory. If it could be shown that extinction is a result of purely physical changes, then one would have no need here of natural selection. Darwin shows that the inhabitants of different regions do not correspond neatly to the different climates. You will not find equivalent life forms just because the conditions are similar.

His gaze roams over the earth, comparing cold regions and warm regions in different zones:

QUOTATION

the first great fact which strikes us is, that neither the similarity nor the dissimilarity of the inhabitants of various regions can be accounted for by their climatal and other physical conditions.

cross reference P: 344; O: 280; ML: 482

If inhabitants are not conditioned by climate, then what are the major influences? Darwin focused future research on the networks of relationships between animals and plants, rather than seeking one grand answer. But the main factor in shaping these relationships in any given place is the possibility or impossibility of further inhabitants arriving from outside the network:

QUOTATION

A second great fact which strikes us in our general review is, that barriers of any kind, or obstacles to free migration, are related in a close and important manner to the differences between the productions of various regions.

cross reference P: 345; O: 281; ML: 483

In particular, Darwin suggests that if the same species is found in two different places, it must at some stage have travelled between them. The only alternative to **migration** is that the same species was created twice, independently. For Darwin, this would be a magical, rather than a scientific explanation, and would root all life forms in a random series of miracles.

QUOTATION

Undoubtedly there are very many cases of extreme difficulty, in understanding how the same species could have migrated from some one point to the several different and isolated points, where now found. Nevertheless, the simplicity of the view that each species was first produced within a single region captivates the mind. He who rejects it, rejects the vera causa *[simple cause] of ordinary generation with subsequent migration, and calls in the agency of a miracle.*

cross reference P: 349; O: 285; ML: 488

He directs future research to the study of patterns of migration. In these chapters, the perspective oscillates between the global and the local. If you query how an animal or plant can have travelled across seemingly impassable spaces, you need only look around the garden. Darwin's gaze sharpens again:

QUOTATION

When a duck suddenly emerges from a pond covered with duck-weed, I have twice seen these little plants adhering to its back.

cross reference **P**: 375; **O**: 311; **ML**: 524

By extension, plants may have travelled across oceans, or to the far side of mountain ranges. True, it is not likely on any given occasion that a plant will be carried over the sea on the foot of a bird, but you then need to bear in mind the vast expanses of time involved. Over such aeons, an isolated instance is extremely plausible and a few such incidents would be enough to spread this plant to a new location: '*the view of occasional means of transport having been largely efficient in the long course of time*' (cross reference **P**: 384; **O**: 320; **ML**: 536).

This is why '*it is an almost universal rule that the endemic productions of islands are related to those of the nearest continent, or of other near islands.*' Here Darwin is returning to his study of the Galapagos islands, all those years before.

In his final chapters, Darwin defends his vision of the planet. Though the tone is polite, and understated, the controversy is more explicit. He turns to the future of the science itself. Previously, natural history has been dominated by the attempt to refine a

> ORIGIN
> **Chapter XIII**
> '*Mutual Affinities of Organic Beings*'
> (**ML**: XIV)

system of perfect classification. Behind that apparently scientific method, there have lurked religious motives:

> QUOTATION
>
> *But many naturalists think that something more is meant by the Natural System; they believe that it reveals the plan of the Creator.*
>
> cross reference P: 399; O: 335; ML: 554

Darwin now feels entitled to dismiss this goal of an absolute system, modelled on an original divine pattern. Through this version of the science, he concludes, 'nothing is added to our knowledge.' The future of natural history lies with the study of development, of dynamic and continuing processes, and not of fixed categories: '*the natural system is founded on descent with modification*' (cross reference **P**: 404; **O**: 340; **ML**: 560).

Biology will be concerned not with ever-better distinctions, but with ever-finer connections. Natural history will become truly a continuing 'history', rather than a mere chart. Instead of making a fetish of their own system, future scientists will explore the ways in which life forms have interacted and changed. Now Darwin draws more directly on his notebooks, bringing to the fore his terminology of 'descent' that lay in the background of his main discussion. He is proposing nothing less than the genealogy of life on earth.

ORIGIN
Chapter XIV
'*Recapitulation and Conclusion*'
(**ML**: XV)

The sense of controversy intensifies still further in the Conclusion, where Darwin famously defines the nature of his great work: '*As this whole volume is one long argument*' (cross reference **P**: 435; **O**: 371; **ML**: 612). Beyond the theory and the mass of data, there is this 'one long argument', for a new way of approaching the natural world, and against the orthodoxy which looked at nature as the expression of a single act of creation or as the planned scheme of a Creator.

Darwin sums up. Significantly, he starts with a review of the gaps and difficulties. His aim is to show that nowhere do these problems provide a basis for refuting the theory. The argument cannot be finally proven on every count; but nowhere has it broken down. Then he brings briefly to mind the key concepts. First there is 'the constantly recurrent struggle for existence' and then 'natural selection' which *'acts solely by accumulating slight, successive, favourable variations'*. As if turning back to his endless list of unused examples, Darwin adds: *'Many other facts are, as it seems to me, explicable on this theory'* (cross reference **P**: 445; **O**: 381; **ML**: 626).

The Conclusion leaves us with a tremendous feeling of knowledge in reserve, and of future research that will deepen the project still further. Darwin then underlines the most controversial claim, though in as low key a way as he can manage. In a typically polite passage, he refers with apparent tolerance to *'our natural unwillingness to admit that one species has given birth to other and distinct species'* (cross reference **P**: 452; **O**: 389; **ML**: 639). But coiled in the sentence is that vivid phrase *'given birth to'*. This is a significant summing up and one which sharpens the 'long argument'. Natural history is now to be the study of how life forms give birth to each other, and not the mapping of the Creator's plan.

The final note is one of modest prophecy, the voice of the founder foreseeing his own achievement surpassed by what he has made possible: *'We can dimly foresee that there will be a considerable revolution in natural history'* (cross reference **P**: 455; **O**: 391; **ML**: 643).

6 Impacts of the *Origin*

THE FIRST IMPACTS

Stephen Jay Gould, a leading authority on Darwin, sums up the importance of the *Origin*: 'Within a decade, he had convinced the thinking world that evolution had occurred' (*Ever Since Darwin*, (1971) pp. 11–12).

Though as Gould says, Darwin held back from using the term, he had launched all the more effectively the theory which became known as 'evolution'. The Darwinian gaze was not only observant and analytic, it was also persuasive. To look at things in this way turned out to be an immensely powerful means of influencing others. This gaze is a deeply contagious way of looking. Once you have seen the world like this, it is hard to step outside the vision. For succeeding generations, the natural world has become an entangled bank, a scene of competition, governed by natural selection, and not a chaotic mingling of chance occurrences, on the one hand, or a place of divine munificence on the other.

There was, however, a time lag in the influence of Darwin's ideas. Gould takes up the story: 'But his own theory of natural selection never achieved much popularity during his lifetime. It did not prevail until the 1940s'. The 'struggle', and the associated idea of conflict, caught on before the even more radical concept of natural selection. Instead of being the source of advantages and adaptations, evolution could appear mainly as the process by which the weak or unfit would be inevitably eliminated. Gould insists that in its original form: '*The essence of Darwin's theory lies in his contention that natural selection is the creative force of evolution – not just the executioner of the unfit.*' But 'struggle' was the first theme to make its impact on the wider imagination.

This time lag reflects another aspect of the Darwinian gaze. What it does see depends on what it also does *not* see. In particular, the Darwinian gaze reveals a world which shows no signs of divine guidance or purpose. Darwin sent an advance copy of the *Origin* to his friend, the Rev. Adam Sedgwick, who was a leading authority on geology as well as a cleric. Sedgwick wrote a famous letter to Darwin on receiving this copy:

> *Tis the crown and glory of organic science that it does through final cause, link material to moral ...* **You have ignored this link;** *and if I do not mistake your meaning, you have done your best in one or two pregnant cases to break it.*
> Adam Sedgwick (1859) in Gillespie, *Genesis and Geology* (1951), p. 217

Here we can feel the threatening quality of Darwin's gaze. It illuminates a world which no longer expresses a higher purpose or source. What is not visible any more, if we take this new viewpoint, is the unity of creation under the benign guidance of a ruling deity. It is only the gaze of such a deity, as in the original Genesis narrative, which keeps all the types firmly in their places, including the human type or species. Sedgwick is particularly anxious about the implied slippage in the status of humanity:

> *Were it possible (which, thank God, it is not) to break it, humanity, in my mind, would suffer a damage that might brutalize it, and sink the human race into a lower grade of degradation than any into which it has fallen since its written records tell of its history.*

In fact, the Darwinian gaze saw – and sees – a world of infinite possibilities, in every minute variation. But for the orthodox, even someone as sympathetic as Sedgwick, this new way of looking appeared merely to be overlooking the most important truth of all. Perhaps this was why the whole theory took so long to break through: it is this new 'scrutinizing' gaze which displaces the previous regard which underwrote each 'kind' and 'saw that it was good'.

Confirmation of the dark side of the impact of the *Origin* can be found widely in the writings of the later nineteenth century. H.G.Wells's *The Time Machine* (1895) is perhaps the blackest of all these Darwinian visions. In this novel, a time traveller discovers a future in which humanity has given rise to two separate species, the Eloi and the Morlocks. The Eloi seem to be living a life of innocent idleness in a pastoral world. Down below ground, the Morlocks appear at first to be their servants. It turns out that the subterranean race is keeping the Eloi like a herd of domestic animals and eating their meat in dark, underground chambers.

Wells's traveller is familiar with the idea of struggle in its Darwinian sense, and he tries to make sense of this situation using that concept. First, he theorizes that the Eloi are leading the empty life that comes after the struggle is over. Then he recognizes the darker truth: there is no end to the struggle for existence. In their way, the Eloi and the Morlocks illustrate Darwin's original metaphorical use of the word 'struggle', which we saw in Chapter III of the *Origin*. Their interaction is a perfect blend of conflict and dependence. Wells's story also cunningly mixes together other Darwinian themes, like the parallel between domestication and natural selection. In a final post-Darwinian irony, humanity itself has become a victim of its own artificial selection. Both the underground Morlocks and the passive Eloi are outcomes of selection by the other side. The Morlocks were meant to be adapted to working-class life; instead, they took over and bred the Eloi to serve as food!

THE CONTINUING INFLUENCE

The ideas of the *Origin* have echoes widely in modern consciousness. For example, the whole understanding of conflict has been coloured by differing interpretations of Darwin's theory. There is a would-be Darwinian streak in many defences of the capitalist free market:

The growth of a large business is merely a survival of the fittest ... This is not an evil tendency in business. It is merely the working out of a law of nature and a law of God.

John D. Rockefeller Jr in Singer, *A Darwinian Left* (1999), p. 11

But , on the other side, Marx saw himself as the Darwin of the social sciences:

Darwin's book is very important and serves me as a natural-scientific basis for the class struggle in history.

Marx letter of 1862 in Singer, *A Darwinian Left* (1999), p. 20

The existentialist Friedrich Nietzsche adds another possibility, in which Darwinian ideas serve a theory of human struggle and transformation:

All creatures hitherto have created something beyond themselves: and do you want to be the ebb of this great tide, and return to the animals rather than overcome man.

Nietzsche, *Thus Spoke Zarathustra* (1883–4)

All of these reactions now feel a long way from the self-critical aspect of Darwin's own argument. None of these grand answers corresponds well with Darwin's own majestic sense of the limits of his own thought:

QUOTATION

It is good to try in our imagination to give any form an advantage over another. Probably in no single case should we know what to do in order to succeed.

cross reference P: 129; O: 65; ML: 106

In the twentieth century, the concept of natural selection had an increasing impact, particularly in conjunction with the rising science of genetics. One side of that influence took the form of **eugenics**, the sometimes sinister science of 'improving' human populations. This approach influenced both the left and the right of the political spectrum, and it coloured the interpretation of Darwin, producing the trend sometimes called 'Social Darwinism'.

Two examples of the direct appropriation of the *Origin* for eugenic arguments can be found in J.B.S. Haldane's *Daedalus* (1924) and Ronald A. Fisher's *The Genetical Theory of Natural Selection* (1930). Haldane tried to mix eugenics with Marxist social design, and a faith in technology. Fisher hoped to help natural selection on its way in improving the human species. But Haldane, in particular, also made an eminent contribution to the development of genetic science.

In our time, genetics has advanced far beyond the limits of that period, and Darwin's legacy has moved on. At the time he wrote the *Origin*, Darwin could know nothing of modern genetics, of course, as it emerged only out of the work of Gregor Mendel later in the nineteenth century. Lacking this dimension, Darwin was unable to explain *how* variations arose or were passed on. But his language continues to be the basis of contemporary genetics, as we can see from a brief quotation from a leading influence on the formation of the modern approach:

> *A gene is being favoured in natural selection if the aggregate of its replicas forms an increasing fraction of the total gene pool.*
>
> Hamilton (1972) in Brown, *The Darwin Wars* (1999), p. 23

What you see – according to this influential argument – when you look at the genetic level is exactly the kind of process that Darwin observed on the entangled bank or in the virgin forest. This is an approach which another leading Darwinian, Richard Dawkins, has called 'the neo-Darwinian modern synthesis', a mixture of new genetic science with prior Darwinian language. This is a new gaze, but still relies fundamentally on the same way of looking as the one introduced by the *Origin*.

Even now, there are deep controversies among scientific interpreters of this great idea. For some, such as Richard Dawkins, natural selection finds its ultimate purpose in telling the story of genetic changes:

> *natural selection must have worked on similar genetic variation in wreaking evolutionary change.*
>
> Dawkins, *The Extended Phenotype* (1982), p. 24

For others, such as Stephen Jay Gould , the focus remains on whole organisms in their environments:

> *…the father of evolutionary theory stood almost alone in insisting that organic change led only to increasing adaptation between organisms and their own environment.*
>
> Gould, *Ever Since Darwin* (1991, reissue from 1977)

Gould is interested in stories of adaptation, successful and unsuccessful, by organisms to environments including, of course, their webs of living relations with other organisms. Dawkins, on the other hand, wants to focus on what he argues are the fundamental transformations which occur at the genetic level and not at the level of the individual organism. This difference has led to debates so heated as to be termed, in one leading contemporary account, by Andrew Brown, 'The Darwin Wars'!

What does the Darwinian story *mean*? Are there moral or social implications? Darwin reflect on his own stories of natural selection:

QUOTATION

there are many unknown laws of correlation and growth, which, when one part of the organisation is modified through variation, and the modifications are accumulated by natural selection for the good of the being, will cause other modifications.

cross reference P: 134; O: 71; ML: 114 (wording varied)

As we have seen, he stresses that the underlying laws are unknown to him. Equally important for his future impact is that little phrase '*for the good of*'. In modern debates, some scientists interpret 'for the good of' in terms of 'the good of the species'. A major figure here is Tinbergen. By contrast, others like Dawkins denounce this interpretation as 'group selectionism'. They insist that the ultimate focus in natural selection must be on individuals and their genes.

Every passage of Darwin's original account has been combed in defence of these rival positions. For example, Darwin remarks on the way parents and offspring fit together:

> QUOTATION
> **Natural selection** *will modify the structure of the young in relation to the parent, and of the parent in relation to the young. In social animals it will adapt the structure of each individual for the benefit of the community.*
>
> cross reference P: 135; O: 72; ML: 115

There are two interpretations of the story. Dawkins sees it in terms of 'exploitative manipulation within the family'. The eminent philosopher, Peter Singer, on the other hand, traces the history of an alternative Darwinism, which sees such passages as evidence of the cooperative dimension of the evolution of species, including the human species.

Darwin foresaw a revolution in natural history. Though he refrained from staking the claim more directly, his *Origin of Species* was itself the origin of that revolution, a transformation in understanding that has affected every other aspect of modern inquiry and debate. Some revolutions honour their founders ceremonially and then forget about them in practice. But we are still living through the Darwinian revolution – and in that continuing upheaval of thought, *The Origin of Species* is sure to yield yet further meanings and insights. We are still acquiring the Darwinian gaze.

REFERENCES AND FURTHER READING

Amigoni, David and Wallace, Jeff (eds), *Darwin's The Origin of Species: New Interdisciplinary Essays* (Manchester University Press, 1995).

Axelrod, Robert, *The Evolution of Co-operation* (Penguin, 1990).

Beer, Gillian, *Darwin's Plots* (Routledge, 1985).

Brown, Andrew, *The Darwin Wars* (Penguin, 1999).

Campbell, John Angus, 'Charles Darwin: Rhetorician of Science' in Nelson, J. Megill, A. and McCloskey, D.N. (eds), *The Rhetoric of the Human Sciences* (University of Wisconsin Press, 1987).

Dawkins, Richard, *The Extended Phenotype* (Oxford University Press, 1982).

Dawkins, Richard, *The Selfish Gene*, second edition (Oxford University Press, 1989).

Dennet, Daniel, *Darwin's Dangerous Idea* (Penguin, 1996).

Desmond, Adrian and Moore, James, *Darwin* (Penguin, 1992).

Dromanraju, Krishna, *Haldane's Daedalus Revisited* (Oxford University Press, 1995).

Gillespie, Charles Coulston, *Genesis and Geology* (Harper & Row, 1951).

Gould, Stephen Jay, *Ever Since Darwin* (Penguin, 1991, reissue).

Gould, Stephen Jay, *Wonderful Life* (Penguin, 1991).

Jones, Steve, *Almost Like A Whale: The Origin of Species Updated* (Doubleday, 1999).

Kevles, Daniel J., *In The Name of Eugenics* (Penguin, 1986).

Singer, Peter, *A Darwinian Left* (Weidenfeld & Nicolson, 1999).

Beyond the three editions of the *Origin*, other references to Darwin:

Darwin, Charles, *The Voyage of the 'Beagle'* (Everyman, 1959).

Porter, Duncan M. and Graham, Peter W. (eds), *The Portable Darwin* (Viking Penguin, 1993).

Other editions used:

Lyell, Charles, *Principles of Geology*, selected and edited by James Secord (Penguin, 1997).

Malthus, Thomas, *An Essay on the Principle of Population,* edited by Anthony Flew (Penguin, 1970).

INDEX

Other related titles

MILL'S *ON LIBERTY* – A BEGINNER'S GUIDE

GEORGE MYERSON

'It's a free country!', we say, when we want to make our own decisions, or express our own ideas. Easily said: but what *really* makes a society free? Written in 1859, John Stuart Mill's *On Liberty* still provides the classic answer. As we enter the third millennium, few of our societies would pass Mill's test of true freedom!

George Myerson's lively text:

* Investigates the background to Mill's *On Liberty*

* Offers a clear and concise summary of the whole book

* Gives close-up explanations of the most important arguments

* Offers a critical appreciation of Mill's contribution to the theory of human liberty.

Other related titles

SMITH'S *WEALTH OF NATIONS* – A BEGINNER'S GUIDE

MARTIN COHEN

Few books have had more impact on the modern world than Smith's celebrated inquiry into the origin of the *Wealth of Nations*. Originally published in the year of the American Declaration of Independence, it rapidly became the most popular book of its time, marking the transition of political society from economic naïvety to financial sophistication.

Martin Cohen's accessible guide to the text:

* Offers a clear and complete summary of the main arguments presented by Smith
* Describes the social and cultural context of the work
* Explains the key concepts of modern economics

Other related titles

NIETZSCHE'S *THUS SPAKE ZARATHUSTRA* – A BEGINNER'S GUIDE

GEORGE MYERSON

Nietzsche's *Thus Spake Zarathustra* is the most controversial book in the history of philosophy. With its many famous lines and statements – including the notorious news that 'God is dead!' – *Thus Spake Zarathustra* has provoked and enthralled generations of readers. Literary masterpiece and angry polemic, manifesto and poem, personal confession and historic prophecy: what will you find in Nietzsche's visionary theory?

George Myerson's lively text:

* Investigates the background of *Thus Spake Zarathustra*
* Offers a clear and concise summary of the whole book
* Gives close-up explanations of the most important arguments
* Focuses for the reader the central concepts such as the superman, redemption and eternal return.